ICE AGE EARTH

The reconstruction of Late Quaternary climate changes draws from a wide range of disciplines. In an area of increasing interest and specialization, *Ice Age Earth* combines a summary of research and a synthesis of Late Quaternary climate history. It focuses on changes in the Earth's geology and climate between the last interglacial and the final melting of the last great ice sheets: between approximately 130,000 and 7,000 years ago.

The book describes the study of ocean sediments and ice cores and considers global circulation models and numerical reconstructions of Late Quaternary climate. Emphasis is placed on regional variability in glacial history, the rapidity of past climate changes and the complexity of environmental responses to the melting of the last ice sheets. Periglacial environments as well as lakes, rivers, volcanic and aeolian activity, crustal and sea level changes are also considered. The book closes with a contextual discussion of Milankovitch insolation theory.

Ice Age Earth provides the first detailed review of global climate change in the Late Quaternary. It should prove essential reading for all students interested in Quaternary environments.

Alastair Dawson is Reader in Quaternary Science in the Department of Geography at Coventry Polytechnic.

A volume in the Routledge *Physical Environment* series.

A VOLUME IN THE
ROUTLEDGE PHYSICAL ENVIRONMENT SERIES
Edited by Keith Richards
University of Cambridge

The Routledge Physical Environment series presents authoritative reviews of significant issues in physical geography and the environmental sciences. The series aims to become a complete text library, covering a range of themes in physical geography and environmental science, including specific processes and environments, environmental change on a variety of scales, policy and management issues, as well as developments in methodology, techniques and philosophy.

Other titles in the series:

WATER RESOURCES IN THE ARID REALM
E. Anderson and C. Agnew

ENVIRONMENTAL HAZARDS
K. Smith

MOUNTAIN WEATHER AND CLIMATE
2nd edition
R.G. Barry

Forthcoming:

THE GEOMORPHOLOGY OF DESERT DUNES
N. Lancaster

GLACIATED LANDSCAPES
M. Sharp

HUMID TROPICAL ENVIRONMENTS AND LANDSCAPES
R. Walsh

SOILS AND ENVIRONMENT
S. Ellis and A. Mellor

PROCESS, ENVIRONMENT AND LANDFORMS
Approaches to Geomorphology
K. Richards

ICE AGE EARTH

Late Quaternary geology and climate

Alastair G. Dawson

London and New York

First published 1992
by Routledge
11 New Fetter Lane, London EC4P 4EE

Simultaneously published in the USA and Canada
by Routledge
a division of Routledge, Chapman and Hall, Inc.
29 West 35th Street, New York, NY 10001

Typeset in Scantext September by
Leaper & Gard Ltd, Bristol
Printed and bound in Great Britain by
Biddles Ltd, Guildford and King's Lynn

British Library Cataloguing in Publication Data
Dawson, A.G. (Alastair George) *1952*–
Ice age earth: late quaternary and climate.
1. Quaternary period
I. Title
551.792

ISBN 0–415–01566–9
ISBN 0–415–01567–7 pbk

Library of Congress Cataloging in Publication Data
has been applied for

For Sue and Gregory

CONTENTS

CONTENTS

ix

PLATES

FIGURES

TABLES

PREFACE

In this book, I have attempted to provide an interdisciplinary account of the geology and climate of the Late Quaternary, drawing on information from a wide range of specialist disciplines. It has been a daunting challenge to write a book that summarises the vast literature on this subject. Almost every week, it seems, exciting new articles appear in journals that deserve attention. I must confess that I have found it almost impossible to have read all of them. Inevitably, I have had to omit important material from the book and in this sense the book is an incomplete account.

My principal objective has been to summarise the most important scientific findings about the nature of Late Quaternary climate changes. This has also included an attempt to provide a literature review in a way that may encourage a greater awareness of the importance of interdisciplinary research. I have endeavoured to create a balanced text that is not biased towards my own interests and covers global rather than regional or local issues. I therefore apologise to those who feel that I have omitted important local information or have presented incorrect interpretations.

I hope that the book will be useful to students of the Quaternary who wish to learn more about climate change and landscape evolution. I hope also that the text will be of particular value to undergraduate students as they endeavour to make sense of seemingly countless and often contradictory academic papers. I hope also that the book will encourage more specialists in different Quaternary research areas to talk to each other and discuss the broader aspects of their research. Most of all, I hope that the book is stimulating and enjoyable to read!

Many people have contributed to the writing of this book. I owe the greatest debt to Dr J.B. Sissons who first introduced me to Quaternary research and whose support and encouragement over the years has been a great strength and inspiration. May I also offer sincere thanks to several people who critically read and improved the quality of individual chapters: Dr C. Firth and Dr J.J. Lowe whose many detailed and constructive criticisms of individual chapters are greatly appreciated; Dr K. Richards whose perceptive comments greatly improved the quality of the text; Professor J. Rose whose suggestions regarding several chapters are also gratefully acknowledged. Constructive criticism regarding individual chapters was also kindly provided by Dr D. Benn; Mr M. Berrisford; Dr C. Chesner; Dr J.J. Clague; Dr C.M. Clapperton; Dr A. Dugmore; Professor R.W. Fairbridge; Dr I. Foster; Dr J. Gordon; Dr D.H. Keen; Dr I. Livingstone; Dr D. Long; Dr S.B. McCann; Professor D.E. Smith and Professor J. Teller.

Grateful thanks are also extended to Mr David Orme who kindly drew the diagrams; Mrs S. Addleton who prepared the Tables; Tristan Palmer and Alison Walters at Routledge, and Neville Hankins for their valuable support, advice and help. Finally thanks to my wife, Sue, for her loyalty and tolerance, her comments on individual chapters and her understanding of my considerable domestic short-comings during the writing of the book.

Alastair Dawson

ACKNOWLEDGEMENTS

Every effort has been made to trace the holders of individual copyright. May I offer my sincere apologies for any omissions that may have occurred. I am grateful to the following for permission to reproduce copyright material:

Longman Group for Figures 4. (Velichko *et al.* 1984), 5.4 (Lowe and Walker 1984), 6.1 and 6.2 (Baulin and Danilova 1984), 6.3 (Washburn 1979, Pewe 1983), 6.4 (Pewe 1983), 7.3 (Smith and Street-Perrott 1983) and Table 8.2 (Baker 1983); Edward Arnold for Figures 6.5 and 9.7 (Washburn 1979); Cambridge University Press for Figure 6.6; Macmillan Magazines Ltd for Figures 2.6 (Johnsen *et al.* 1972), 3.5 (Street-Perrott and Perrott 1990), 3.6 (Thiede 1974), 9.8 (Petit *et al.* 1990), 10.3 (Ninkovich *et al.* 1978) and 12.8 (Fairbanks 1989); Academic Press for Figures 5.1 (Grosswald 1981), 5.3 (Eronen 1983), 7.2 (Hamilton 1982), 10.2 (Drexler *et al.* 1980), 10.5 (Ruddiman and Glover 1975), 10.7 (Sancetta *et al.* 1973), 11.4B (Eronen 1983), 13.5 (Ruddiman *et al.* 1980); The Minister of Supply and Services Canada for Figures 4.4, 5.5, 5.6, 5.7, 5.8 and 5.9 (Dyke and Prest 1987); Oxford University Press for Figures 7.5 (Street and Grove 1979), 9.5 and 9.6 (Goudie 1983); Allen & Unwin Publishers for Figures 7.4 (Nicholson and Flohn 1980 and Bradley 1985) and 13.2; John Wiley & Sons for Figures 3.1, 4.3 and 13.1 (Skinner and Porter 1987), 3.3, 3.4, 4.5, 9.2 and 9.3 (Flint 1971), 11.4A (Balling 1980), 12.5 (Donner 1980) and Tables 4.4 (Denton and Hughes 1980), 12.1 and 12.6 (Mörner 1980); Basil Blackwell Ltd for Figure 12.1 (Mörner 1987); Scientific American for Figure 4.7 (Broecker and Denton 1990); Institute of British Geographers for Figures 11.3 and 12.7 (Andrews 1970); Pergamon Press for Figures 2.2 and 12.4 (Shackleton 1987), 3.2 (Kutzbach and Wright 1985), 4.2 (Behre 1989 and Larsen and Sejrup 1990), 4.6 and 13.2 (Andrews *et al.* 1986), 11.1 (Bowen 1981), and Tables 2.1 (Bowen 1981) and 4.2 (Larsen and Sejrup 1990); Geological Society of America for Figures 3.7 (Lozano and Hays 1976), 9.4 (Kukla *et al.* 1988), 10.6 (Bogaard and Schmincke 1985); Ellis Horwood Ltd for Figure 9.1 (Catt 1988); Routledge, Chapman and Hall (Methuen) for Table 8.1 (Chorley *et al.* 1984); the American Geophysical Union for Figures 2.5B (Lorius *et al.* 1988) and 11.7 (Crittenden 1963); Universitetsforlaget, Oslo for Figure 5.2 (Mangerud *et al.* 1979); Springer Verlag for Figure 10.3C (Fisher and Schmincke 1984); Balkema, Rotterdam for

Figures 8.2 (Adamson and Williams 1980) and 8.3 (Talbot 1980) and North-Holland Publishing Company for Figure 10.4 (Gow and Williamson 1971).

Many individuals have kindly given permission for the inclusion of diagrams and Plates in this book and I am indebted to them for permission to use their copyright. In particular my thanks go to Dr M. Acreman for Plate 12; Professor A.L. Berger for Figures 13.2, 13.3, 13.4, 13.5, 13.6; Sue Dawson for Plate 1; Dr D. Drewry for Figure 11.2; Dr A. Dugmore for Figure 10.1; Professor C. Embleton for Plate 8; The Geological Survey of Canada for Plate 14; Dr C. Green and Dr D.F.M. MacGregor for Figure 8.1; Dr M. Eronen and H. Olander for Figure 7.1; Dr J. Karte for Figure 6.6; Dr C. Lorius for Figure 2.5B; Dr D. Munro for Plate 16; Mr A. Newton and Mr J. Finlay for Plate 13; Dr Y. Ota and University of Tokyo Press for Figure 12.3; Dr O. Van de Plassche for Figure 12.2; Professor S.C. Porter for Plate 15; Dr K. Richards for Plate 9; Mr M. Punkari for Plate 3; Dr M. Seddon for Plate 6; Dr H.-P. Sejrup for Plate 2; Professor I. Smalley for Plate 10; Dr J. Teller for Figure 11.6 and Plate 5; Dr D. Thomas for Plate 11; Professor A.L. Washburn for Figures 6.5 and 9.7; Dr A. Werrity for Plate 4; Dr D. Vere for the cover photograph.

1

INTRODUCTION

INTRODUCTION

The purpose of this book is to describe evidence for fluctuations in the Earth's climate that took place during the Late Quaternary – a time interval corresponding approximately to the last 130,000 years. It attempts to provide an account of the most important geological and geomorphological changes that occurred as a result of global and regional changes in climate. It also endeavours to describe the patterns of climate change and explain why they took place.

The field evidence for past variations in the Earth's climate has been used by scientists interested in predicting future climates. In particular, it has enabled the development of advanced computer models of global climate change (general circulation models) in which past changes in global atmospheric and oceanic circulation have been numerically simulated. Recently, for example, studies of Antarctic ice cores have provided new information for the models on long-term variations of atmospheric CO_2 and CH_4 that can usefully inform discussion of modern global warming caused by Greenhouse gases. Similarly, recent reconstructions of environmental changes that took place during the melting of the last ice sheets have identified past periods of time when climate change took place extremely rapidly. This information is of great value in the study of modern global climate and its vulnerability to sudden changes. It is also of great value with regard to our ability to predict the regional impact of future climate changes.

The book opens with a review of past climate changes based on studies of ocean sediments and ice cores. Thereafter, a review is given of computer models of the general circulation of the Earth's atmosphere and oceans during the last ice age. This is accompanied by a comparative analysis of modern (interglacial) and ice age global circulation. Considerable attention is also given to the history of glaciation since the end of the last interglacial. Emphasis is placed on the regional variability in glacial history and the rapidity at which such changes appear to have taken place. A separate account is given on the melting history of the last ice sheets. This discussion also stresses the rapidity of past climate changes and the complexity of the environmental responses to widespread deglaciation.

A systematic account is given in later chapters of geomorphological changes that occurred in areas that were largely unaffected by glacier and ice sheet development. Thus the consideration of ice age periglacial environments discusses ways in which studies of Late Quaternary periglacial phenomena have contributed to our understanding of past climate changes. Similarly, a discussion is given on the information that studies of lakes, bogs and mires have provided on Late Quaternary climate variations. An account is also given on the evolution of Late Quaternary rivers as well as the response of aeolian activity to past climate fluctuations. There is also a discussion of the importance of volcanic ash layers in the dating of Late Quaternary sediments and a consideration of possible relationships between the largest volcanic eruptions of the Late Quaternary and periods of widespread global cooling. The nature of Late Quaternary crustal movements is also described and there is a detailed consideration of past sea level changes.

In the final chapter, there is an introductory description of Milankovitch insolation theory together with an account of reconstructed patterns of Late Quaternary Milankovitch insolation variations. The different types of geological evidence for past climate changes are summarised in this chapter and consideration is given to possible relationships between past climate changes, based on geological evidence, and Milankovitch insolation variations.

Any discussion of the timing and duration of Late Quaternary global climate change must risk stepping through a minefield of controversy. Of course, the reconstruction of past climate changes is dependent upon the application of reliable dating techniques and an understanding of the limitations of these methods. It should be stressed that this book does not consider these aspects of palaeoenvironmental reconstruction since there are already many other excellent textbooks that consider these subjects (e.g. Lowe and Walker 1984; Bradley 1985). Throughout the book, a conscious attempt has been made to use published ages for particular environmental changes that are considered the most reliable. For simplicity and ease of reading, standard error ranges for individual dates are omitted and the reader is referred to the published papers for this information. It is also assumed that the conventional radiocarbon timescale does not depart markedly from that based on sidereal years.

The information provided in the following chapters demonstrates that studies of Quaternary palaeoenvironments are highly interdisciplinary in character but there is still a great need for scientific specialists to be aware of the broader context of their research. Bradley (1985) has pointed out the danger and the challenge inherent in such developments. He argued that the danger is that in the future there will be limited interdisciplinary understanding and collaboration. He stated that the challenge is to avoid this pitfall and to develop an awareness of the advances of the various methodological developments that are being made. To that must be added the need to learn more about the past climate and environmental changes themselves. It is hoped that this book may contribute something towards this goal.

TIMESCALES AND TERMINOLOGY

General time subdivisions

The Quaternary period, the later part of which is discussed in this book, is often subdivided into two epochs or series – the Pleistocene and Holocene. The Quaternary period is often subdivided into sections – Early, Middle and Late. The boundary between the Early and Middle Quaternary is usually defined by the Matuyama–Brunhes geomagnetic polarity reversal, considered here to have occurred near 790,000 BP (Johnson 1982). The boundary between the Middle and Late Quaternary is regarded as equivalent to the beginning of oxygen isotope substage 5e that represents the warmest phase of the last interglacial. Here, the age of this boundary defined from marine oxygen isotope stratigraphy is considered to be 130,000 BP (Martinson *et al.* 1987) (Figure 1.1).

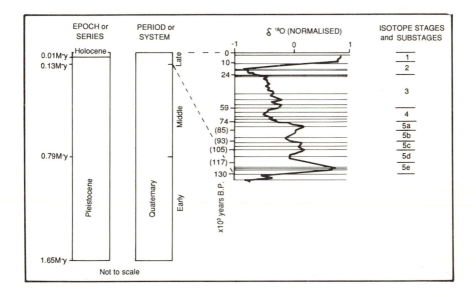

Figure 1.1 Generalised time subdivisions of the Quaternary period (after Goudie 1983). The Late Quaternary is additionally subdivided into oxygen isotope stages (after Martinson *et al.* 1987).

Time subdivisions of the Late Quaternary

In most parts of the world, the time interval between the start of oxygen isotope substage 5e (130,000 BP) and 10,000 BP is assigned a formal chronostratigraphic name[1] (e.g. Weichselian: NW Europe; Devensian: British Isles; Wurm:

central Europe; Valdai: European USSR; Wisconsin: North America). With the exception of the chronostratigraphic subdivisions for the USA and Canada, these time intervals are subdivided into three and accorded the prefixes Early, Middle and Late (e.g. Early, Middle and Late Weichselian). In recent years, these time subdivisions have been linked to the marine oxygen isotope chronology and this has been used to assign age ranges to the respective time intervals (Figure 1.1). For example, in Scandinavia the Early Weichselian is considered to represent the time interval between marine oxygen isotope substages 5d and 5a, the Middle Weichselian to isotope stages 4 and 3 and the Late Weischselian to isotope stage 2. The boundary between the Pleistocene and Holocene (isotope stage 1) is arbitrary but is generally regarded as having occurred near 10,000 BP. Discussion of environmental changes that took place during the present (Holocene) interglacial is here confined to the Early Holocene – ending at approximately 7,000 BP. This time was chosen since it coincides broadly with the final disappearance of the remnants of the last great ice sheets in the northern hemisphere and the return of most glacial meltwater into the world's oceans. It also coincides approximately with the beginning of significant human impact on the environment and the reader is referred to Roberts (1989) for a more complete discussion of Holocene environmental changes that took place after 7,000 BP. For convenience therefore, the term Late Quaternary is not used in the text in its strictest sense since it does not consider environmental changes that took place after circa 7,000 BP.

Oxygen isotope chronology

A study of the various academic papers on Late Quaternary environmental changes will leave the reader very confused about the ages of the respective stages and substages outlined in oxygen isotope stratigraphy. Recently, a revision of the most accurate stage and substage ages has been undertaken by Martinson *et al.* (1987) (Figure 1.1). This was accomplished by comparing the oxygen isotope chronology based on studies of deep sea sediments (see Chapter 2) with the pattern of variations in the Earth's orbit geometry according to Milankovitch theory (see Chapter 13). Martinson *et al.* (1987) used this concept of 'orbital tuning' to develop a high-resolution timescale for the Late Quaternary. Owing to its relatively recent publication, the oxygen isotope chronology used by Martinson *et al.* (1987) does not coincide with the age intervals for the Late Quaternary that are presently used in the USA and Canada. Furthermore, the age subdivisions that are used in Canada and the USA also differ from each other (see Table 4.1).

NOTES

1   Chronostratigraphy is one of several categories of stratigraphic subdivision. It refers to the classification of stratigraphic units according to age. Other categories are lithostra-

tigraphy (the classification of local sediment units or rock successions according to changes in lithology), biostratigraphy (classification of sediment units according to variations in fossil content) and morphostratigraphy (the classification of landforms according to their relative ages) (see Lowe and Walker 1984).

2

OCEAN SEDIMENTS AND ICE CORES

OXYGEN ISOTOPE VARIATIONS IN OCEAN SEDIMENTS

Our understanding of Quaternary climate changes has been revolutionised by the study of sediments that have accumulated on the floors of the world's oceans. This is because ocean floor sediments generally provide a continuous record of sedimentation and only on rare occasions have they been disturbed by episodes of erosion. Such exceptions occur in areas where submarine landsliding and turbidity current activity have taken place. Sediments over much of the ocean floor consist predominantly of the skeletal remains of deposited calcareous and siliceous micro-organisms that have settled out of the water column. Most calcareous tests are dissolved below about 4 km water depth (the calcite compensation depth (CCD)) due to dissolution in the undersaturated water (Bradley 1985). As a result, most ocean cores used in palaeoenvironmental studies are taken from water depths shallower than approximately 4 km. Investigations of the undissolved species present within ocean sediment cores may therefore be used to reconstruct former patterns of environmental change.

Evidence of the former marine conditions under which the calcareous micro-organisms lived can be determined by the analysis of the stable isotope ratios of the oxygen in the carbonate skeletal remains. Thus the relative abundance of the oxygen isotopes in the tests of calcareous micro-organisms can be related to the abundance of these isotopes in the ocean water at the time the carbonates were produced. In this respect, the calcareous microfossils from ocean sediments that have been most commonly employed in oxygen isotope analysis are those of protozoans known as foraminifera. Planktonic foraminifera have provided valuable information on former ocean surface conditions while benthonic species have provided information on former deep water ocean environments.

Modern oxygen isotope variations

The stable oxygen isotope composition of the carbonates produced by the micro-organisms and finally deposited upon death on the ocean floor is partly a function of the isotopic composition of the ocean water during the life span of the

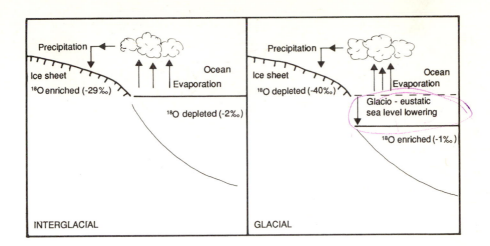

Figure 2.1 Schematic diagram showing in relative terms the respective enrichments and depletions of oxygen isotopes 16 and 18 during glacial and interglacial ages. The oxygen isotope curves for both ice cores and ocean cores are normally measured as ^{18}O variations calculated with reference to the standard isotopic value of the PDB-1 reference belemnite from the Cretaceous Pee Dee Formation of South Carolina. Typical values of ^{18}O for glacial and interglacial ages are shown as values per millilitre. Note that the measured ^{18}O values for marine foraminifera are very small.

micro-organism and also of the temperature of the water in which the foraminifera lived. During the evaporation of sea water the different isotopes of oxygen in water molecules (^{16}O, ^{17}O and ^{18}O) are released into the atmosphere by a process known as isotopic fractionation (Figure 2.1). The process is based on the fact that since the rate of water evaporation varies according to the density (and hence atomic weight) of the respective oxygen isotopes, there is a preferential evaporation of the lighter oxygen isotopes. Changes in the isotopic composition of sea water are mostly dependent on variations in ^{18}O: in general ^{16}O is passive in the cycle of isotope transfer. Oxygen isotope ratios of foraminifera carbonate are also influenced greatly by the water depth and the temperature of the water in which carbonate precipitation is taking place. Thus, for every 1°C fall in water temperature there is a relative enrichment within faunal shell debris of about 0.02 parts per millilitre (‰) in the proportion of ^{18}O with respect to ^{16}O.

The ratios of oxygen isotopes are calculated as relative deviations (^{18}O per millilitre) from the mean ratios of a standard value (Shackleton and Opdyke 1973) as,

$$^{18}\text{O} = 1000 \left((^{18}\text{O}/^{16}\text{O})\text{sample}/(^{18}\text{O}/^{16}\text{O})\text{standard} \right) - 1$$

Consequently, the deviations are calculated as positive or negative values relative to a standard zero value. For example, an ^{18}O value of $-4‰$ corresponds to a sample that is 4 parts per millilitre deficient in ^{18}O relative to the standard value (see Figure 2.1). The measurements that have been made of modern oxygen isotope ratios of precipitation (expressed as deviations from a standard) exhibit marked variations between continent and ocean owing to the influence of fractionation and water temperature variations. For example, in the northern North Atlantic Ocean, the influence of the Gulf Stream leads to the occurrence of much lower depletions of ^{18}O (typically between $-5‰$ and $-10‰$) than would otherwise be the case. By contrast, the values measured for the central area of the Greenland ice sheet are as low as $-37‰$ while values for the (unglaciated) Canadian Arctic are typically of the order of $-20‰$.

Oxygen isotope variations during ice ages

During periods of glaciation, ocean surface waters, as a result of their lowered temperatures, are enriched in the denser ^{18}O isotope. Conversely, snow precipitation deposited on ice sheets during a glacial age is depleted in ^{18}O. During periods of interglacial warmth the reverse process takes place with large quantities of ^{18}O removed from the oceans. The oxygen isotope ratios of foraminifera that lived in ice age oceans were also largely influenced by water temperature. Water temperature variations during glacial periods may have been significantly lower near the surfaces of certain oceans and hence may have altered the oxygen isotope ratios of planktonic foraminifera. Benthonic foraminifera, on the other hand, exist in bottom waters whose temperatures are close to freezing point. It is thought that the temperature of oceanic bottom water has not changed appreciably between glacial and inter-glacial periods. Thus, in the reconstruction of Late Quaternary environments, benthonic species may often contain an oxygen isotope signature that is only slightly influenced by past ocean temperature variations and which is mostly attributable to variations in the isotopic composition of sea water.

Most oxygen isotope studies have been based on the analysis of planktonic foraminifera and have sought to distinguish the proportion of the oxygen isotope signal that is attributable to water temperature changes from that caused by variations in the isotopic composition of sea water. In order to do this, it is necessary to calculate from best estimates the former isotopic composition of sea water during a glacial age. First, one must estimate the total volume of ice sheets and glaciers during the last glacial maximum and thereby calculate the average oxygen isotopic composition of this ice. Second, the volume of ocean water during the last glacial maximum has to be determined, and, using the reconstructed isotopic content for the ice sheets, the likely isotopic composition of the ocean waters can then be estimated. It is the calculation of these values that has posed great problems during the last three decades and has led to radically different interpretations of the oxygen isotope record provided by the ocean core data. These are well illustrated in Table 2.1 where the calculation of water volume

8

Table 2.1 Estimates of the effect of Quaternary ice sheets on the isotopic composition of the oceans

Source	Water volume change × 10⁶ km³	¹⁸O of ice (per mil)	Effect on oceans ¹⁸O per mil (PDB-1)
Emiliani (1955)	58	−15	0.4
Craig (1965)	100	−17	1.5
Olausson (1965)	65	−35	1.7
Shackleton (1967)			1.2–1.6
Dansgaard and Tauber (1969)	47	−30	1.2

After Bowen (1978).

equivalent stored in the last great ice sheets ranges from 47 to 100 million km³. Accordingly there have been considerable variations in the estimates of the isotopic composition of the world's oceans during ice ages.

Interpretation of oxygen isotope curves

Emiliani (1955) calculated from published sources the volume of the last ice sheets and thus attempted to estimate their theoretical isotopic composition as well as their effects on the isotopic composition of the worlds oceans. He argued that this difference was equivalent to 0.4 parts per millilitre between glacial and interglacial ocean waters (Table 2.1). However, since his measurement from ocean cores of the total isotopic variation between a glacial and an interglacial was about 2 parts per millilitre, he concluded that approximately 30 per cent of this difference was due to isotopic changes in the composition of sea water while 70 per cent of the difference was attributable to temperature changes in the surface ocean waters in which the planktonic foraminifera lived. Emiliani therefore maintained that the curves of oxygen isotope variation derived from ocean cores were principally an expression of former ocean temperature variations.

Later studies challenged Emiliani's value of 0.4 parts per millilitre. Thus Shackleton (1967) proposed that the true figure lay between 1.4 and 1.6 per millilitre while Dansgaard and Tauber (1969) proposed that 1.2 per millilitre was a more accurate estimate (Table 2.1). If correct, these calculations imply that most of the measured isotopic variations is attributable to changes in the isotopic composition of ocean water rather than to changes in ocean temperature. Furthermore, since the principal factor controlling the isotopic composition of ocean water is the volume of the continental ice sheets, the oxygen isotope curves predominantly represent an approximate record of global ice volume fluctuations and, by inference, global glacio-eustatic sea level changes.

The principal area of disagreement between Emiliani (1955) and Dansgaard and Tauber (1969) was the estimated ¹⁸O content of continental ice sheets during the last glaciation. Emiliani believed that the value was −15‰ for ice in Green-land and Antarctica and −9‰ for the Laurentide and Fennoscandian ice sheets. He used an estimate of the volume of ocean water locked up in these Late Quaternary ice sheets and calculated that the transfer of this water with an isotopic content of −15‰ would have resulted in a corresponding reduction in isotopic variation of −0.4‰ in the remaining ocean waters. Dansgaard and Tauber proposed instead that the true value for ice was nearer −30‰ (Table 2.1).

There are numerous sources of uncertainty in the results of oxygen isotope analyses. These are summarised only briefly here and the reader is referred to Bowen (1978: 64–9), Bradley (1985: 171–223) and Lowe and Walker (1984: 144–54) for more detailed information and discussion. Suffice to state that the oxygen isotope curve from ocean sediments must be considered as a useful approximation rather than an exact curve of palaeoglaciation. Moreover, since it is representative of an aggregate ice volume, the curve cannot be used as a surrogate for the sequence of glaciation in any individual region of the globe.

More recently, isotopic studies of benthonic foraminifera have provided a more accurate measure of palaeoglaciation. This is because benthonic water masses appear not to have been subject to major temperature fluctuations between glacial and interglacial periods (see below). However, benthonic foraminifera are relatively rare and very few detailed studies have been under-taken (see Dansgaard 1984: 209–13; Shackleton 1987). Nevertheless, these and other studies indicate isotopic variations in benthonic foraminifera that appear to have resulted principally from changes in the isotopic composition of sea water and which hence are believed to provide a more accurate record of palaeoglacia-tion. Inspection of Figure 2.2 thus provides valuable information on the trend of global ice volume changes over the last 130,000 years.

Oxygen isotope analyses have now been undertaken on many cores from all of the world's oceans (Bradley 1985). Most of these investigations have produced similar oxygen isotope curves for each of the major ocean basins. The oxygen isotope fluctuations, as might be expected, are approximately synchronous throughout the world. The most widely quoted curve is that obtained from core V28-238 from the Solomon Plateau region of the equatorial Pacific (1°N, 160°E) (Shackleton and Opdyke 1973) (Figure 2.2). The respective isotope stages deduced from this core are numbered according to a scheme outlined by Emiliani (1955, 1961, 1966). In this, the warmer periods (interglacials and interstadials) are allocated odd numbers and the colder (glacial) periods are assigned even numbers. Note that the isotopic differences between glacial and interglacial periods are relatively small (approximately 1 part per millilitre relative to the standard).

The oxygen isotope curves show that a remarkable sequence of environmental changes took place during the Late Quaternary. First, they indicate that there have been numerous alternating periods of warm and cold climate. Second, they show that many of these fluctuations have been extremely rapid (e.g. the changes

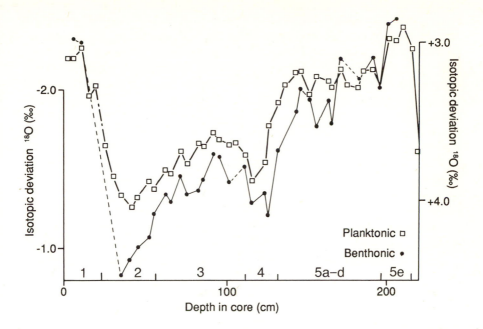

Figure 2.2 Oxygen isotope curves from core V28-238 from Solomon Plateau, equatorial Pacific, based on benthonic and planktonic foraminifera (after Shackleton 1987). The oxygen isotope stage numbers are also shown (after Shackleton and Opdyke 1973).

from glacial stages 2 and 6 to interglacial stages 1 and 5 respectively). Third, the changes in global ice volume appear also to have been of considerable magnitude. Finally, they demonstrate that during the Late Quaternary there have been relatively few periods of time when global ice volumes have been as low as they have been during the present interglacial (isotope stage 1) (Figure 2.2).

The oxygen isotope curve for stages 5 to (e.g. core V28-238) based on analyses of planktonic foraminifera may be compared with that derived from analysis of the benthonic foraminifera (Figure 2.2). Both curves exhibit similar signatures and imply that during the last ice age in this area of the Pacific, both surface and bottom water temperatures did not differ markedly from those of the present. Consequently both curves appear to portray a reasonably accurate chronology of global ice volume fluctuations during the Late Quaternary and Early Holocene.

The role of ocean bottom water (deep water)

Detailed analysis of benthonic foraminifera in ocean cores has indicated that in certain ocean areas the oxygen isotope stratigraphy shows that slight changes in the temperature of bottom ocean water may have occurred during the last ice age. For example, Shackleton and Opdyke (1973) concluded from comparison of the planktonic and benthonic isotope records from Pacific core V28-238 that there had been a decrease in the temperature of bottom water in the equatorial Pacific of 0.5°C during isotope stages 5–2. They attributed the production of this cold water to meltwater discharge from the Laurentide ice sheet (Figure 2.2) (see Chapter 4).

The temperature of ocean bottom waters appears also to have fluctuated considerably in the North Atlantic during the Late Quaternary. For example, Duplessy *et al.* (1980) have argued that there was a decrease of 1.3°C in the temperature of North Atlantic bottom water during isotope stages 4, 3 and 2 and have also related this to meltwater input from the northern hemisphere ice sheets. Duplessy *et al.* (1980) consider that during the last glaciation there was a change not only in the rate of production of bottom water but also in the location where it was produced. At present, most North Atlantic bottom water is produced in the Norwegian and Labrador Seas where water of relatively high salinity is convected and upwelled towards the surface of the ocean where it is cooled (Figure 3.5). The cooling of the saline ocean waters results in an increase in water density that, in turn, causes the water to sink to the floor of the North Atlantic. This process is greatly accelerated by the seasonal development of sea ice that, through the expulsion of salts, increases the salinity and density of the underlying waters causing them to sink. In the Antarctic Circumpolar Ocean, the seasonal development of sea ice also assists the production of bottom water by similar processes. Much of the deep water in the world's oceans originates in the North Atlantic. From this area, deep water is dispersed into the South Atlantic Ocean where it joins the Antarctic Circumpolar Current and thereafter spreads into the major ocean areas of the southern hemisphere (Broecker and Denton 1990a,b).

Bottom water temperature fluctuations during the Late Quaternary are still poorly understood. However, it is clear that the patterns of surface ocean circulation and temperature change during the Late Quaternary have been profoundly influenced by changes in bottom water production and circulation. The reconstruction of past bottom water conditions are necessary since temperature variations in bottom water during the Late Quaternary may have affected the true meaning of individual oxygen isotope curves. For example, oxygen isotope curves based on benthonic foraminifera data may not always provide an extremely accurate record of changes in the isotopic composition of sea water since a small part of the isotope signal may be attributable to bottom water temperature changes.

Limitations of oxygen isotope analysis of ocean sediment cores

In ocean areas where the rate of sedimentation has been relatively low during the Late Quaternary, it has proved difficult to obtain accurate records of the oxygen isotope stratigraphy. Indeed, even in areas where former sedimentation rates were high, the effects of bioturbation caused by the burrowing of bottom-dwelling organisms in the surface sea floor sediments have greatly constrained the age-accuracy of the data. As a result, the most precise time resolution that one might hope to obtain from a core is approximately 1,000 years.

Oxygen isotope analysis depends upon the secretion of carbonates by foraminifera occurring in isotopic equilibrium with the surrounding ocean waters. Some species of foraminifera do not do this. Many calculations of oxygen isotope ratios have been undertaken on mixed species of foraminiferal tests and are thus of uncertain accuracy. Additional problems are posed by certain foraminiferal species that secrete carbonate only during specific months of the year and those that migrate through the water column during secretion. Moreover, many foraminiferal tests are dissolved as they sink towards the ocean floor to water depths below the calcite compensation depth.

On a different scale, relative displacement of foraminifera may occur through submarine landslides and turbidity currents (Ericson *et al.* 1956; Jansen *et al.* 1987) and great care must therefore be taken in the interpretation of the oxygen isotope stratigraphy for individual regions since submarine mass movement events may provide a distorted stratigraphy.

Late Quaternary oxygen isotope stratigraphy from ocean sediments

The numerous cores that have been analysed for their oxygen isotope stratigraphy from the different oceans of the world have yielded, with exceptions, similar patterns of oxygen isotope fluctuations for the last 130,000 years. This is not surprising since the oxygen isotope signature is dependent on the changing isotopic composition of sea water over time. This is most prevalent in studies using benthonic foraminifera since isotopic changes due to water temperature fluctuations are never very large. However, oxygen isotope investigations using planktonic foraminifera have often provided evidence of major temperature fluctuations of ocean surface water between glacial and interglacial periods. In general, these have been characteristic of ocean surfaces where subpolar waters have been replaced by polar waters during periods of glaciation on adjacent continents. This is well illustrated from studies of ocean cores from the North Atlantic and neighbouring Labrador and Norwegian Seas where pronounced sea surface temperature changes appear to have taken place during the Late Quaternary (Belanger 1982; Kellogg 1986; de Vernal and Hillaire-Marcel 1987; Larsen and Sejrup 1990). By contrast, studies of planktonic foraminifera from cores from equatorial ocean areas generally indicate small sea surface temperature changes between glacial and interglacial periods.

The North Atlantic Ocean, Norwegian Sea and Labrador Sea

Sancetta *et al.* (1973) provided detailed information of sea surface water temperature changes for the North Atlantic during the Late Quaternary. Their results show clearly that dramatic temperature decreases took place after the last interglacial during isotope substages 5d and 5b. Their data also show the even more severe and sustained temperature decline during the transition between isotope stages 5 and 4. Each of these climatic deteriorations appears to have been considerable, with surface ocean temperature decreases of the order of 7°C.

Detailed research by Ruddiman *et al.* (1980) has also provided valuable information on Late Quaternary oxygen isotope variations for numerous cores from the North Atlantic. These studies have used benthonic foraminifera as indicators of global ice volume fluctuations together with planktonic foraminifera as indicators of surface water conditions. Their analyses have provided information on variations in (a) sea surface temperature, (b) sea surface salinity, (c) the supply of ice-rafted detritus, and (d) bottom water conditions for the North Atlantic over the last 130,000 years.

Ruddiman *et al.* (1980) demonstrate very clearly that there was a major period of global ice sheet growth during the transition between oxygen isotope stages 5 and 4 at 75,000 BP. Their sea surface temperature data led to the remarkable conclusion that most of the ice accumulation on neighbouring continents took place while most of the North Atlantic between 40°N and 60°N remained nearly as warm as present with salinities similar to present (Figure 2.3). It was concluded that 'warm' ocean surfaces coupled with a decline in solar radiation during this period provided the necessary conditions for rapid northern hemisphere glaciation. Furthermore, Ruddiman *et al.* (1980) observed that there was a relatively low supply of ice-rafted detritus during the initial period of ice sheet growth, a factor that they attributed to dominantly land-based ice sheets that were incapable of producing significant quantities of icebergs. The relatively warm sea surface temperatures calculated by Ruddiman *et al.* (1980) for the North Atlantic during the isotope stage 5/4 transition are in striking contrast with those calculated for the same area during the culmination of the last glacial maximum (stage 2) at 18,000 BP (Ruddiman and McIntyre 1981b) (Figure 2.3). This difference implies that northern hemisphere ice accumulation was produced during the stage 5/4 and 2 intervals by quite different processes. During the isotope stage 5/4 interval, the juxtaposition of 'cold' continents and 'warm' oceans led to ice sheet build-up. During isotope stage 2, ice sheet growth on 'cold' continents' was fed by moisture from 'cold' oceans.

A more controversial interpretation by some authors is that major sea surface cooling and the development of a seasonal sea ice cover took place in the Norwegian Sea initially during isotope substage 5d and lasted until the end of isotope stage 2 (Duplessy *et al.* 1975; Kellogg 1976; Kellogg *et al.* 1978) (Figure 2.4). Kellogg *et al.* (1978) concluded that the period between isotope substages 5e and 5a was characterised by a dramatic decline in sea surface temperature that began at the end of substage 5e. More recent analyses indicate that there may have been seasonal open water until the end of isotope substage 5a while brief periods of ocean warming may also have occurred at the beginning and at the end of stage 3 (Belanger 1982; Larsen and Sejrup 1990). The occurrence of sustained cold conditions during the Late Quaternary has also been proposed by Kellogg (1986) for the eastern Labrador Sea. Kellogg (1986) has argued that this

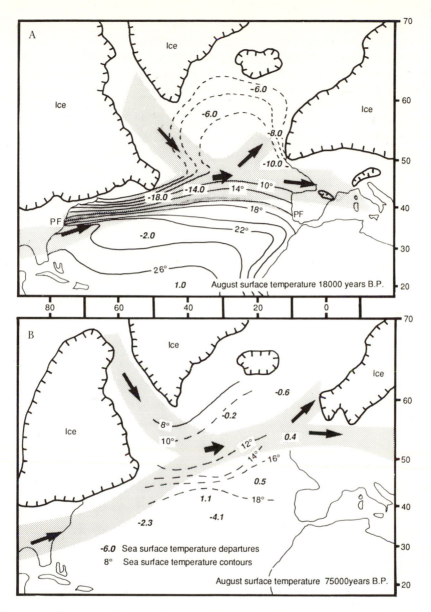

Figure 2.3 Estimated North Atlantic August sea surface temperatures and selected temperature anomalies for 75,000 BP and 18,000 years BP. The ice cover for both periods is schematically illustrated. Note that at 75,000 BP the build-up of ice sheets on 'cold' continents was associated with an abundant moisture supply from a 'warm' ocean. By contrast at 18,000 BP, 'cold' continents lay adjacent to a 'cold' North Atlantic Ocean. The principal jet stream tracks are also indicated (after Kutzbach and Wright 1985) while in A, the position of the polar oceanic front is shown (PF) (after Ruddiman *et al.* 1980; Price 1983).

15

ocean area was characterised by a year-round sea ice cover for most of the period between isotope substage 5a and the transition between stages 2 and 1.

The unresolved question of whether open water was present or absent at the end of substage 5a in the North Atlantic, Norwegian Sea and Labrador Sea areas is a critical factor in the model of rapid ice build-up at 75,000 BP as proposed by Ruddiman *et al.* (1980) since temperate oceanic conditions are necessary to provide the moisture needed for a source of snow precipitation and, by inference, ice accretion on neighbouring landmasses. Detailed studies of ocean floor sediments from the Labrador Sea area by de Vernal and Hillaire-Marcel (1987) show clearly the influence of ocean water conditions on the timing of ice accretion over eastern Canada and Greenland during the Late Quaternary. Their oxygen isotope analyses differ from those obtained from low latitude areas in that the signal is greatly distorted by fluctuations in the input of glacial meltwater during the growth and decay of the Laurentide and Greenland ice sheets. Nevertheless, their research indicates that, with the exception of isotope stage 2, there appears to have been a continuous penetration of the North Atlantic Drift into the Labrador Sea throughout the Late Quaternary and hence there is likely to have been a suitable source of moisture for snow precipitation (Figure 2.4).

OXYGEN ISOTOPE ANALYSIS OF ICE CORES

Background

During periods of glaciation, the evaporation of moisture from ocean surfaces leads to the eventual precipitation of oxygen isotopes within snow, and their preservation in glacier ice (Figure 2.1). Consequently, global climatic fluctuations can be detected through the study of $^{18}O/^{16}O$ ratios in ice cores. In theory, the isotopic variations in ^{18}O should bear an approximate inverse relationship to fluctuations observed for ocean core isotope records for the same time periods.

To date, over 80 cores of over 100 m length have been sampled from ice sheets, ice shelves and glaciers in both hemispheres (Bradley 1985). Of these, the most significant are those taken through the Antarctic and Greenland ice sheets. The principal problem associated with interpretation of ice cores is the need to know the former patterns of flow within the ice. Clearly, if significant ice flow has taken place in the area of the core, the core is of little value. Therefore, attempts have been made to recover ice cores from the centres of individual domes within ice sheets (Figure 2.5). It is also necessary to retrieve cores from areas where there has been little surface elevation change in the ice, since such changes may have affected the measured ^{18}O values.

Inspection of an idealised model of an ice sheet profile (Figure 2.5) shows that in a single-domed ice sheet in the area beneath the ice divide where radial flow is minimal, significant plastic deformation of ice takes place. Consequently, the time resolution in ice cores becomes progressively poorer with increasing depth

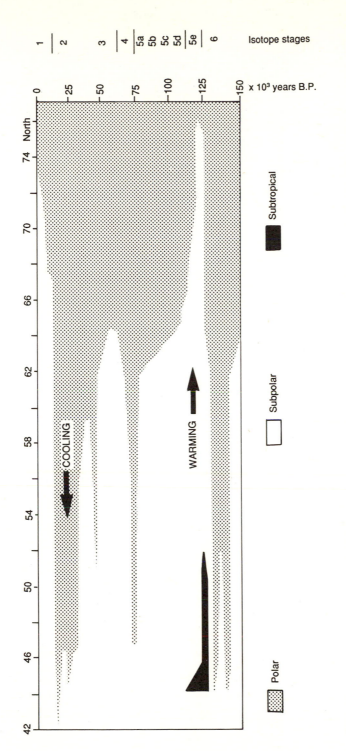

Figure 2.4 Faunal and floral assemblage variations in the Norwegian Sea and northern North Atlantic for the past 150,000 years based on planktonic foraminifera and coccolith data (after Belanger 1982). The variations indicate that this region was affected by several periods of profound oceanographic change during the Late Quaternary (after Belanger 1982).

A

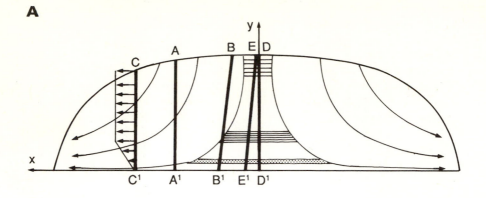

B

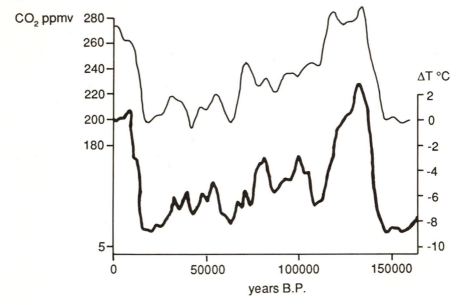

Figure 2.5 A. Schematic diagram to illustrate flow patterns within an idealised ice sheet and the limitations associated with the dating of ice cores. In the diagram showing a single-domed ice sheet, a series of ice cores sampled at D would present the most complete record of past climate changes although here age resolution would decrease with increasing depth due to the effects of ice compression. Vertical cores taken from the surface at sites A or C would present an incomplete record of climate change due to former ice flow (shown by parallel arrows). Cores sampled obliquely at sites B and E would also be of little value. B. Curve of CO_2 variations over last 150,000 years from the Vostok ice core, Antarctica (values in p.p.m.v.) (upper diagram) and a smoothed version of the Vostok isotope temperature record (lower diagram) (after Lorius *et al.* 1988).

and ice age. Furthermore, an adjustment of the ^{18}O value has to be made to allow for the downward movement of ice beneath the ice divide as well as the vertical accretion of ice during ice sheet growth. It is rare, therefore, to find ice core records that extend beyond the last 100,000 years.

It is extremely difficult to provide an accurate dating chronology for ice cores. Detailed ice core measurements have been made of long-term variations in ^{18}O. These have enabled the development of a relative isotope chronology for the Late Quaternary that can be compared with those derived from studies of ocean sediments. Theoretical models of ice sheet flow have also been used to predict the calculation of ice age at depth. In addition, reference layers have often been used to define ice age (e.g. particular layers of volcanic ash, loess dust and other microparticles). Recently, the use of the cosmogenic isotope of beryllium (^{10}Be) has enabled estimates to be made of former variations in snow precipitation (Raisbeck et al. 1987; Jouzel et al. 1989). It is generally agreed, however, that the composite use of dating methods provides the most accurate ice core chronologies. Very accurate timescales, however, only exist for about the last 10,000 years. For example, it has proved possible to date the age of ice younger than 8,000 years by counting summer layers of ice that exhibit seasonally high isotope values (Johnsen et al. 1972). For ice older than this, age–depth determinations become increasingly dependent on theroetical models of ice sheet flow.

Interpretation of ice core data

Despite the difficulties in the calculation of age–depth relationships from ice cores, many of the cores recovered from both the northern and southern hemispheres exhibit remarkable similarities in their isotopic variations. In the northern hemisphere, the well-known core from Camp Century, Greenland, exhibits clearly the dramatic ice sheet build-up that took place during isotope substage 5d (Dansgaard et al. 1982) (Figure 2.6). Glacial conditions (i.e. high ^{18}O depletions) are a characteristic feature after 80,000 BP until the end of isotope stage 2. More temperate conditions are indicated for the Holocene. However, the Camp Century isotope data have never been converted to a temperature scale due to the fact that the area from which the core was taken may have been subject to considerable changes in ice surface elevation during the last 100,000 years (Dansgaard 1984).

Cores taken from Antarctica at Byrd Station and at Vostok (Lorius et al. 1979) display similarities not only between each other but also with the Camp Century core and one from Devon Island in the Canadian Arctic (Koerner 1977; Bradley 1985). In each case the curves of isotopic variation may be altered by using different values for inferred ice accumulation rates, ice flow divergence and ice sheet thickness increase rates during the periods of ice accumulation. The assumption of a constant ice accumulation rate over time is a particular weakness. Indeed, with regard to the Camp Century core, Dansgaard et al. (1982) concluded that the accumulation rates may have been irregular, having

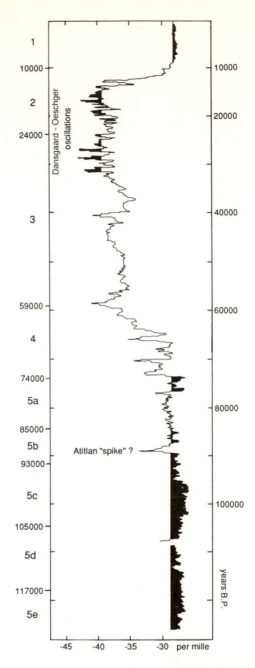

Figure 2.6 The ^{18}O record from Camp Century, Greenland. Prominent Dansgaard–Oeschger oscillations are shown as well as the isotopic 'spike' that may or may not be related to the Atitlan volcanic eruption (see Chapter 10) (after Johnsen *et al.* 1972). Oxygen isotope stages are also shown (left).

been higher during periods of ice sheet growth (e.g. during isotope stage 4) and lower during periods of peak glaciation (e.g. stage 2).

Spikes of excessively high ^{18}O depletions occur in several of the Greenland ice cores for the time period between 55,000 and 25,000 BP and indicate pulses of global cooling during oxygen isotope stage 3 (Figure 2.6). Broecker and Denton (1990b) have decribed these phenomena as Dansgaard–Oeschger oscillations. They observed that these events have durations of approximately 1,000 years and rise-and-fall times on the order of about a century. Their occurrence implies that, during isotope stage 3, the northern hemisphere was subject to a series of extreme fluctuations in climate with many severe periods of cold conditions. However, similar oscillations are not evident in Antarctic ice cores and it is not yet clear why this should be so. One possibility is that the oscillations signify climate changes that only affected the northern hemisphere (Broecker and Denton 1990b).

The analysis of ice cores has also provided valuable information regarding long-term changes in snow precipitation and variations in atmospheric CO_2. For example, variations in the concentration of the cosmogenic isotope of beryllium (^{10}Be) have been measured from the Vostok ice core, Antarctica (Raisbeck et al. 1987). Since it is assumed that ^{10}Be is deposited at a constant rate over time, changes in ^{10}Be in the Vostok ice core have been considered indicative of long-term changes in the amount of snow precipitation. Studies of ^{10}Be variations indicate that the rate of snow precipitation increased by a factor of two during full glacial conditions and decreased markedly (lower snowfall) during the last interglacial to values similar to present.

Analysis of the Vostok ice core has also provided valuable information regarding CO_2 variations during the Late Quaternary. The CO_2, sampled from air bubbles trapped within the ice, exhibits a striking correspondence with the oxygen isotope record (Figure 2.5) (Lorius et al. 1988). For example, strong Greenhouse warming appears to have taken place during the last interglacial while the CO_2 concentrations were extremely low during periods of glaciation. Lorius et al. (1988) have observed that there is a close correspondence between the CO_2 variations and inferred temperature changes for the Late Quaternary. Lorius et al. observed that during the time period between 125,000 and 75,000 BP the changes in CO_2 concentration clearly lag behind the temperature record. During later periods, however, there is almost simultaneous variation between the CO_2 and temperature records. It is difficult at present to determine the role that CO_2 atmospheric variations may have exerted on Late Quaternary climate changes. Much more research remains to be undertaken on this intriguing topic before satisfactory answers can be provided.

INFERRED PATTERNS OF GLOBAL
ENVIRONMENTAL CHANGE

Global ice volume and sea level fluctuations

Shackleton and Opdyke (1973) calculated a probable sea level curve for the last 130,000 years based on the oxygen isotope data (see Chapter 12). Their premise was that a measured variation of 0.1‰ in ^{18}O is approximately equivalent to a 10 m change in global sea level. Expressed in a different manner, these data also provide information on rates of continental ice sheet growth and decay although they cannot provide any information on where the growth and decay of individual ice sheets took place. Thus, Shackleton and Opdyke (1973) considered that there was a lower global ice volume during isotope substage 5e than during the present interglacial. The isotope curves also indicate that major ice accumulation took place at 115,000 BP (the transition between isotope substages 5e and 5d) with a secondary phase of ice build-up during isotope substage 5b. By contrast, periods of ice volume reduction appear to have occurred during isotope substages 5c and 5a. The curves demonstrate that a major and rapid period of ice build-up took place at or slightly after 75,000 BP during the transition between isotope stages 5 and 4 (Ruddiman *et al.* 1980). A more recent revision of global glacio-eustatic fluctuations during the Late Quaternary has been suggested by Shackleton (1987) (Figure 12.4). Shackleton's (1987) glacio-eustatic sea level curve uses both benthonic and planktonic oxygen isotope data and is probably the most accurate indicator, so far, of Late Quaternary global sea level fluctuations.

Shackleton has estimated that approximately 50 per cent of the ice volume accumulation over continental areas by 18,000 BP had been completed by isotope substage 5d. Furthermore, most of this ice accumulation, perhaps between 75 per cent and 90 per cent of it, had been completed by isotope stage 4. These data would therefore appear to suggest that the major northern hemisphere ice sheets remained essentially intact between the beginning of stage 4 (75,000 BP) until the last glacial maximum at 18,000 BP. The geological evidence, however, appears to suggest that this pattern of events was considerably more complex (see Chapter 4).

Inferred trends in Late Quaternary climatic change

Despite the limitations of oxygen isotope analysis on ice and ocean sediment cores, comparison of the isotope variations derived from the long ice cores and studies of both planktonic and benthonic foraminifera show strikingly similar trends. For example, the onset of cooler conditions at the end of the last interglacial is clearly defined. Similarly the progressive yet quasi-cyclical decline towards colder conditions and the culmination of major ice build-up at circa 75,000 BP are clear features of most profiles. The culmination of the last glacial maximum at 18,000 BP is also well defined in both the ice core and ocean core curves. It is tempting to dwell on the apparent correspondence between the timings of the ^{18}O enrichments and depletions in both the ocean and ice cores. However, there are many assumptions and much more research is needed before the problems of calibration and correlation are adequately resolved.

The patterns of former environmental change provided by ice cores for the Late Quaternary are much more detailed than those of ocean cores. Analysis of the isotopic variations in the ice cores points to long-term climatic changes upon which are superimposed numerous sharp climatic oscillations. For example, it is well known that the Younger Dryas climatic deterioration believed by many to have occurred between 11,000 and 10,000 BP resulted in significant ice advance across NW Europe (Lowe and Gray 1980). However, this event appears to be represented in the Greenland Camp Century core by an ^{18}O oscillation of only minor proportions and certainly no greater than many others which preceded it (Figure 2.6) (Broecker *et al.* 1988). The ice cores also appear to indicate that the last 90,000 years have been punctuated by numerous climatic deteriorations whose magnitudes were greater than that conventionally associated with the Younger Dryas (see Figure 2.6). The overall pattern that the oxygen isotope curves exhibit for the Late Quaternary is one characterised by incessant switches between global cooling and warming. Unfortunately, we are not yet able to tell, except at the most elementary level, where and when these changes took place. Despite the pessimism of this statement, we should not ignore the tantalising glimpses into the past that oxygen isotope analyses of ice and ocean cores have so far provided.

3

ICE AGE PALAEOCLIMATES AND COMPUTER SIMULATIONS

INTRODUCTION

The reconstructions that have been made of the last ice age have provided much information on past climatic conditions. The most unified attempt has been in the CLIMAP Project (CLIMAP Project Members 1976) where detailed reconstructions were made for the Earth's climate for 18,000 BP. More recently, several attempts have been made to develop numerical models of the patterns of global oceanic and atmospheric circulation that existed during the last ice age. These general circulation models (GCMs) represent simulations, based on geological evidence, of the response of the atmosphere to inferred distributions of sea surface temperature, the extent and altitude of former ice sheets, the former distribution of lakes, etc. The GCMs are then tested by the use of other climate parameters not used in the model (e.g. land-based estimates of former temperatures). During recent years, numerous GCMs have been developed, each different from each other and each providing different results. A particular difficulty has been that the geological evidence of past environmental conditions in different areas of the world is highly controversial. Thus the accuracy of individual GCMs is only as reliable as the geological data that are used in the construction of the respective models.

In the following section, a discussion is given on some of the principal general circulation models that have been developed. Thereafter, the most important regional differences between the climate of the last ice age and the present climate are outlined. Some of the characteristics of ice age climate for particular regions are based largely on geological evidence. In other instances, the GCMs have provided the most valuable information regarding past ice age conditions.

ICE AGE GENERAL CIRCULATION MODELS (GCMs)

The culmination of the CLIMAP project resulted in the production of detailed maps of global land and sea surface temperatures for 18,000 BP (CLIMAP 1976) (Figure 3.1). The publication of this information enabled the preparation of computer simulations of global ice age atmospheric and oceanic circulation

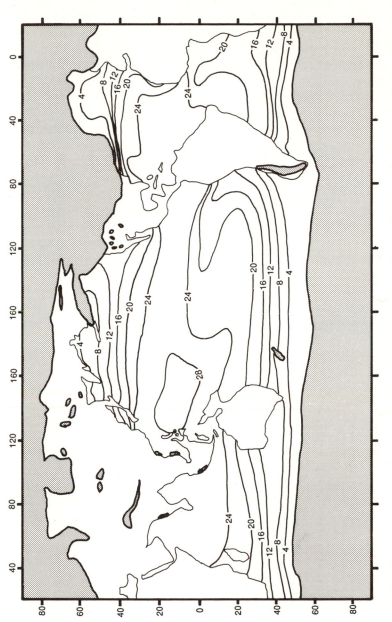

Figure 3.1 CLIMAP map showing reconstructed August sea surface temperatures during the last glacial maximum. Cold polar water reached far south of its present limit in the North Atlantic while plumes of cold water also flowed westwards across the equatorial Pacific and Atlantic (after Skinner and Porter 1987). In several equatorial ocean areas, sea surface temperatures were similar or slightly higher than present. Shaded areas depict high latitude regions occupied by glacier ice and sea ice.

(Gates 1976a,b). The global distribution of ice for this period is, of course, a fundamental element in the model yet is not known with certainty (Denton and Hughes 1981). Accordingly, the CLIMAP project members identified a number of regions where the 18,000 BP glacial limits are controversial (CLIMAP Project Members 1976). These included uncertainty over the extent of the Antarctic ice sheet, the scale of grounded ice cover in the Barents Sea, the distribution of ice shelves in the Arctic Ocean, and the margins of the Cordilleran and Laurentide ice sheets. Gates used the available CLIMAP data for a July (only) 18,000 BP GCM reconstruction. He noted that the simulated ice age evaporation and precipitation were reduced to approximately 15 per cent of their present July values and observed that July air temperatures may have been as much as 10–15°C lower than present across the large continental areas south of the Laurentide and Eurasian ice sheets. Gates (1976a,b) also concluded that ice age atmospheric circulation was dominated by a strengthened zonal (E–W) air flow and a weakening of meridional (N–S) circulation.

In a later study, Manabe and Hahn (1977) concluded that the lowering of land surface temperatures proposed by Gates was essentially correct but noted that for certain regions the estimated temperature lowering may have been underestimated. For example, their model suggested that temperature lowering over northern Alaska was as low as −20°C in contrast to the value of −4°C calculated by Gates (1976a,b).

In a more recent and extremely detailed study, Kutzbach and Wright (1985) attempted to reconstruct global climatic circulation for different time periods between 18,000 BP and the present (Figure 3.2). The boundary conditions used in their model included orbitally induced Milankovitch variations in solar radiation (< 1 per cent) and changes in atmospheric CO_2 content (to 200 p.p.m.). Kutzbach and Wright (1985) and Berger (1979) showed that the calculated Milankovitch Earth orbital parameters for the last glacial maximum were very similar to the present. For example, at 18,000 BP, the Earth had a similar axial tilt (23.44°) while the eccentricity of its orbit (0.0194) was similar to the present value of 0.0167. One difference was that perihelion (when the Earth is closest to the Sun) occurred near the autumn equinox rather than during January (see Chapter 13). Kutzbach and Wright (1985) estimated that, at 18,000 BP as a result of these changes, the value of net incoming solar radiation may even have been 3–4 per cent higher than present during March–May although there was a corresponding lowering between September and November. They also estimated that between January and July, incoming solar radiation was little different from present. It is clear, therefore, that the scale of world glaciation at 18,000 BP was greatly influenced by preceding global environmental changes and that the last great ice sheets greatly influenced the nature of the glacial-age global atmospheric and oceanic circulation systems (see Chapter 13).

In its overall scope, the Kutzbach and Wright (1985) model is confined to northern hemisphere fluctuations and therefore cannot include any influence on global atmospheric circulation exerted by the Antarctic ice sheet and its related

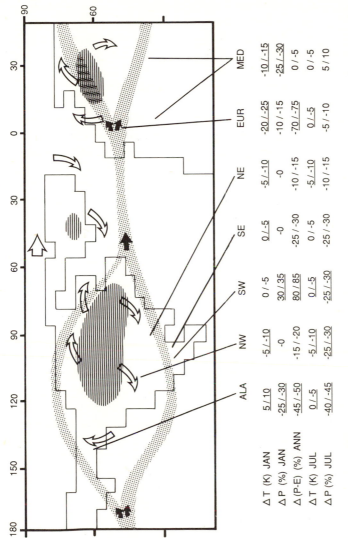

	ALA	NW	SW	SE	NE	EUR	MED
ΔT (K) JAN	5/10	-5/-10	0/-5	0/-5	-5/-10	-20/-25	-10/-15
ΔP (%) JAN	-25/-30	-0	30/35	-0	-0	-10/-15	-25/-30
Δ(P-E) (%) ANN	-45/-50	-15/-20	80/85	-25/-30	-10/-15	-70/-75	0/-5
ΔT (K) JUL	0/-5	-5/-10	0/-5	0/-5	-5/-10	0/-5	0/-5
ΔP (%) JUL	-40/-45	-25/-30	-25/-30	-25/-30	-10/-15	-5/-10	5/10

Figure 3.2 Generalised pattern of northern hemisphere circulation for January 18,000 BP. Surface winds are shown by open arrows and jet stream air flow by continuous stippled area and solid arrows. The principal ice sheet areas are also shown (shaded). Average values of the difference between ice age and modern temperature (K), precipitation (%) and precipitation minus evaporation (%) are shown for Alaska (ALA), northwest USA (NW), southwest USA (SW), southeast USA (SE), northeast USA (NE), Europe (EUR) and the Mediterranean (MED). The values are expressed to the nearest interval of 5 so that, for example, a temperature difference of −7°C is shown as −5/−10. The underlined values are those that are the most statistically significant. Note the pattern of split jet stream flow around both the Eurasian and Laurentide ice sheets (after Kutzbach and Wright 1985).

sea ice cover. Despite this important limitation, Kutzbach and Wright attempted to simulate the January and July patterns of temperature, precipitation and wind for 18,000 BP. The boundary conditions for the model were based on CLIMAP data.

Their model shows that a more zonal southerly jet stream flowed south of the Laurentide ice sheet and thereafter crossed the North Atlantic in the vicinity of the polar oceanic front. It was shown that, during winter, a northern branch of the North American jet stream brought cold air into the North Atlantic via the corridor between the Laurentide and Greenland ice sheets. Kutzbach and Wright stressed that the climatology of North America was dominated by anticyclonic flow around the Laurentide ice sheet, the surface temperature of which was typically 20–30°C lower than present. Thus, the mid latitude cyclones were displaced southwards so that in southwestern USA, SW winds in conjunction with increased precipitation, lower temperatures and increased evaporation provided the conditions necessary for the development of so-called 'pluvial' lakes (see Chapter 7).

This supply of precipitation over southwestern USA may also have been aided by an abundant supply of moisture from the eastern Pacific due to similar ocean surface temperatures to the present although this process may have been countered by enhanced oceanic upwelling. By contrast, the GCM reconstructions for southeastern USA at 18,000 BP suggest greater summer aridity while, in winter, an eastward jet displacement would have led to an intensification of a SW airflow. Kutzbach and Wright (1985) argued that the displacement of the jet stream would also have led to reduced precipitation and greater aridity in the northern Mediterranean and that temperatures south of the ice sheets may have been much lower both in January (by 20–30°(!)) and in July (by over 5°).

The pattern of jet stream flow around the Laurentide ice sheet has been simulated by Kutzbach and Wright (1985), Broccoli and Manabe (1987) and Kutzbach and Guetter (1986). Each describe a 'split' upper tropospheric airflow as having occurred around both the Eurasian and Laurentide ice sheets. The models also depict three distinct anticyclones located over the Laurentide, Greenland and Eurasian ice sheets as well as a broad zone of increased precipitation that extended across the mid latitudes from southwestern USA to western Europe. In all of the computer simulations, the reconstructed values for former precipitation and evaporation provide in their differences a useful indication of former run-off and hence give some indication of ice age lake levels (Figure 3.2). These values, which bear little relationship to the reconstructed values of precipitation, appear to indicate high run-off in southwestern USA (hence high lake levels) yet low run-off in southeastern USA (low lake levels) (Figure 3.2). This is extremely important in any consideration of the so-called pluvial lakes since it demonstrates that lakes and rivers in different regions of the middle and low latitudes are likely to have fluctuated in a very complex manner (see Chapters 7 and 8).

A separate general circulation model for the last glacial maximum in the tropics was undertaken by Manabe and Hahn (1977). Their simulation for a

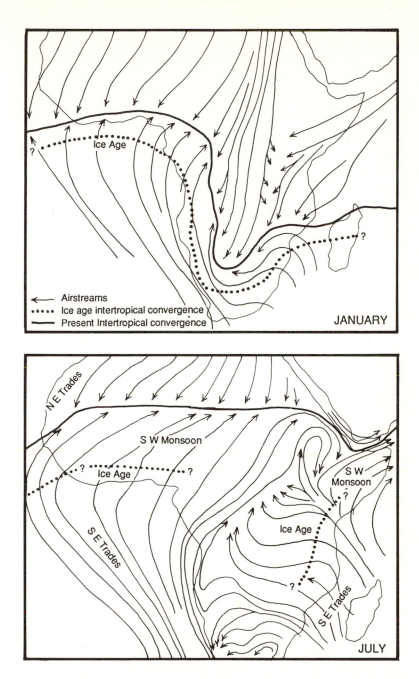

Figure 3.3 Mean zones of convergence and air streams over equatorial and southern Africa in January and July. Hypothetical positions of the Intertropical Convergence Zone (ITCZ) at 18,000 BP are also shown (adapted from Flint 1971).

glacial summer used data similar to those employed by Gates (1976a,b) and showed that tropical continental areas during this period were subject to considerably more extensive aridity than present. In part, this appears to have been due to an increase in global surface albedo which reduced the amount of solar radiation available for evaporation. However, it was also due to generally lower temperatures that further reduced the overall amount of evaporation (Manabe and Hahn 1977). Manabe and Hahn noted that the simulation depicted very low precipitation over Venezuela and tropical Africa, a trend that is consistent with the displacement of the ITCZ and related monsoon precipitation (Figures 3.3 and 3.4). They also concluded that the relatively strong Hadley cells of subtropical high pressure over tropical continental areas may have caused enhanced air outflow from these areas and thus promoted enhanced tropical ice age aridity.

A general theme arising from the general circulation models is that areas south of the major ice sheets during the last glacial maximum were relatively arid as well as having been characterised by reduced temperatures. Elsewhere, in areas where the temperature gradients in the lower troposphere are considered to have been much steeper (e.g. at the polar oceanic and atmospheric fronts), the models invariably depict greatly enhanced jet stream flow. Another trend evident in the simulations is that in areas affected by the polar atmospheric front, the positions

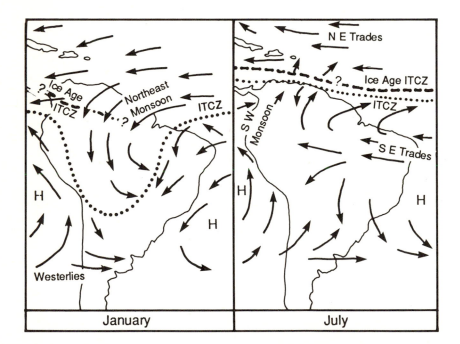

Figure 3.4 Mean positions of the Intertropical Convergence Zone and air streams over South America during January and July and possible ITCZ positions at 18,000 BP (adapted from Flint 1971).

of the mid latitude cyclones (and hence the areas of highest precipitation) tend to occur parallel to the jet stream axis and usually along the SW–NE quadrant of jet flow in the zone where considerable upper tropospheric air divergence is counter-balanced by enhanced air convergence and cyclone development in the lower troposphere. Thus there were two principal environments that provided conditions suitable for the rapid delivery of moisture from the oceans to the ice sheets. One area incorporated those areas of enhanced cyclone production beneath individual SW–NE-trending limbs of the mid latitude jet streams along the polar atmospheric and oceanic fronts. The second source comprised those ocean areas located beneath merging jet streams in the lee of major ice masses where cyclone production was incessant. The most important such area was in the western North Atlantic in the lee of the Laurentide ice sheet.

ICE AGE PALAEOCLIMATES IN THE NORTHERN HEMISPHERE

The development of general circulation models relies upon the use of certain boundary conditions. Some of the atmospheric boundary conditions may be considered as constant values (e.g. the equations of (horizontal) air motion, the thermodynamic energy equation, the hydrostatic equation and continuity equations for atmospheric mass and water vapour (Gates 1976a,b)). However, geological boundary conditions are subject to a great deal of uncertainty. For example, there is considerable disagreement about the extent and elevation of the major ice sheets that existed during the last glacial maximum. Not surprisingly therefore, geological evidence can often provide us with primary information about past ice age climatic conditions for particular regions. The use of general circulation models enables the global simulation of past ice age climates. However, the accuracy of these model predictions of past climates is only as reliable as the geological boundary conditions that are included in the model. In the following sections, use is made of both primary geological information and GCM simulations to reconstruct past ice age climatic conditions for different regions of the world.

Middle and high latitudes

The CLIMAP reconstructions of the Earth's climate for 18,000 BP depict an Earth which, despite a changed geography of ice, land and water, was characterised by patterns of global atmospheric and oceanic circulation analogous in many respects to the present global circulation. In continental areas of the northern hemisphere, the presence of large ice sheets resulted in the development of permanent high pressure cells due to the cooling by the ice of the overlying air. In addition, the ice sheets, which were locally over 3,000 m in thickness, presented major topographic barriers to upper tropospheric jet stream flow. As a consequence, permanent anticyclones developed separately over the Laurentide,

Eurasian and Greenland ice sheets. The clockwise circulation of anticyclonic surface airflow extended far beyond the ice sheets and played a large part in the creation of the extensive loess plains of central Europe, Asia and North America.

During the last glacial maximum, dry anticyclonic conditions were also a characteristic of the Arctic Ocean since a surface sea ice cover prevented moisture evaporation from ocean surfaces. In addition, the presence of extensive sea ice led to the cooling of the surface air and hence promoted air subsidence and stability. One result of these processes was that the higher latitudes of the northern hemisphere were extremely arid during the last glacial maximum since the presence of a sea ice cover over the Arctic Ocean resulted in substantially decreased evaporation of moisture. Another consequence was that permanent high pressure over the Eurasian ice sheet and adjacent non-glaciated regions would have prevented monsoonal circulation over SE Asia. Under these circumstances, the middle and low latitude jet streams are likely to have been subject to a complete reorganisation. One consequence of these changes may have been a split jet stream flow along both the northern and southern margins of the Eurasian and Laurentide ice sheets and around the Tibetan Plateau (Figure 3.2). For example, the GCM of Kutzbach and Wright (1985) clearly shows the diversion of jet stream flow around both the northern and southern flanks of the Laurentide ice sheet and their convergence over the Labrador Sea and western North Atlantic (Kutzbach and Wright 1985).

The southward extension of permanent anticyclones also caused a southward shift of the mid latitude cyclones since the polar atmospheric front was located south of the ice sheet margins. Under such circumstances the principal supply of precipitation to the northern hemisphere ice sheets would have been along their southern margins – thus the ice sheets may frequently have been advancing in the south while undergoing retreat at higher latitudes.

In the continental areas of the middle and high latitudes, where there was insufficient moisture for glacier development and ice sheet growth, the enhanced annual negative heat budget at the ground surface resulted in the aggradation of permafrost (see Chapter 6). In particular, large areas of NE Siberia as well as areas of northern Alaska, northern Greenland and Spitzbergen were characterised by arid permafrost conditions. Permafrost aggradation also took place in the cold tundra environments on continental interior regions located south of the Laurentide and Eurasian ice sheets.

Oceans

North Atlantic and North Pacific Oceans

During the last glacial maximum, the displacement of the polar atmospheric front south of the Laurentide and Eurasian ice sheets was accompanied by similar displacements over the North Atlantic and Northern Pacific Oceans. In both ocean areas, marked changes in the condition of both bottom and surface waters took place. The most significant changes were associated with the development of sea ice. In both the North Atlantic and Northern Pacific during winter months, sea ice extended as far south as 40–45°N (CLIMAP 1976). During summer, a considerable northward retreat of the sea ice margin took place and, in the North Atlantic, may have resulted in the southern edge of permanent pack ice being located near 60°N (CLIMAP 1976). The considerable sea ice cover, by drastically reducing moisture evaporation and by cooling the overlying air, resulted in the southward extension of high pressure.

In the North Atlantic and North Pacific Oceans, the mixing of cold dense polar waters and warmer less dense subtropical waters led to the development of a polar oceanic front whose position was defined by a very steep temperature gradient. As a result, the displaced positions of the polar atmospheric and oceanic fronts led to a corresponding southward displacement of mid latitude cyclones. One area of particular importance in the production of cyclones, and hence of precipitation, was the western North Atlantic in the lee of the Laurentide ice sheet. According to the GCM model predictions of Kutzbach and Guetter (1986), this area appears to have been characterised by the merging of a 'split' jet and was hence an area subject to enhanced air convergence and cyclone development. This may have been of immense importance since most snow precipitation to the eastern sector of the Laurentide ice sheet may have been supplied from this area.

The rate of mixing of polar and subtropical water in the North Atlantic during the last glaciation is also considered to have profoundly influenced global climate due to a marked reduction in the formation of deep water in the North Atlantic (NADW) (Broecker and Denton 1990a,b) (Figure 3.5). Broecker and Denton (1990a,b) proposed that, during the last glaciation, there was a marked increase in the salinity stratification of the North Atlantic due to an increase in the amount of surface glacial meltwaters entering the ocean. The effect of this change was a substantial reduction in the rate of NADW formation in the North Atlantic as well as a decrease in the vertical mixing of the world's oceans that led, in turn, to changes in surface climate (see Chapter 7).

Arctic Ocean

The global lowering of sea level during the last glacial maximum also had a profound influence upon oceanic circulation in the Arctic Ocean. A major consequence was the formation of a Bering land bridge between NW Alaska and NE Siberia that confined water inflow into the Arctic Ocean to the area of the northern North Atlantic. In this region, the thickness of North Atlantic water entering the Arctic was also decreased by approximately 120 m due to global sea level lowering. Thus, there was a reduction in the exchange of heat between the Arctic and the North Atlantic. In addition, the supply of fresh water into the Arctic Ocean was increased due to increased meltwater run-off. This, in turn, increased the salinity stratification near the ocean surface and promoted the development of sea ice.

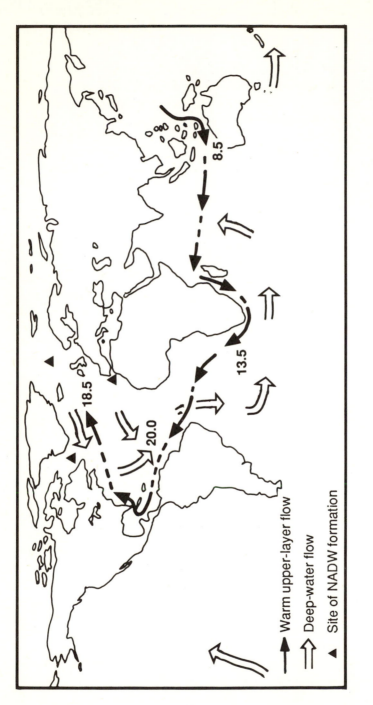

Figure 3.5 Global thermohaline circulation cell associated with North Atlantic deep water (NADW) production. Solid arrows show the inferred warm water route for return flow of upper-layer water to the North Atlantic. The suggested volume fluxes expressed as Sverdrup units (Sv) are based on a uniform upwelling of NADW with a production rate of 20 Sv (based on Street–Perrott and Perrott 1990). One Sverdrup unit is equivalent to $1 \times 10^6 \ \mathrm{m}^3 \ \mathrm{s}^{-1}$.

Warm upper-layer flow
Deep-water flow
▲ Site of NADW formation

8.5
13.5
18.5
20.0

Figure 3.6 Winter sea surface temperatures (°C) in the Mediterranean at 18,000 BP (after Thiede 1974). Black areas denote land areas exposed due to sea level lowering.

The extent and nature of ice cover over the Arctic Ocean during the last glacial maximum is, however, not known with certainty. One view is that the Arctic Ocean was covered by an extensive floating ice shelf that bordered neighbouring ice sheets and which formed a virtually continuous cover across the region. A different view, and one that corresponds with the CLIMAP reconstructions, considers that the Arctic Ocean had an almost complete cover of sea ice. Although this difference of view is of fundamental importance in the reconstruction of former ice sheets, it is not important from a climatological point of view. This is because the presence of an ice cover over the surface of the Arctic Ocean, irrespective of its nature, would have favoured the development of permanent high pressure and anticyclonic conditions.

Mediterranean Sea

During the last glacial period in the Mediterranean Basin, sea level lowering of 120m resulted in the separation of the eastern and western Mediterranean Basins which were connected by only a narrow channel located between Tunisia and Sicily (Figure 3.6). In addition, the reduced water depths (to 230m) in the Straits of Gibraltar resulted in a diminished flux of water between the North Atlantic and Mediterranean. Thiede (1974) attempted to reconstruct the Mediterranean palaeoceanography for 18,000 BP using planktonic foraminiferal analysis (Figure 3.6). The inferred February reconstructions for sea surface temperature ranged between 7°C between southern Spain and Morocco and 18°C in the eastern Mediterranean (compared with present temperatures of near 22°C and 26°C respectively). Thiede (1974) also drew attention to a major influx into the Mediterranean throughout the year of cool surface waters from the Aegean Sea. The source of this cold water was most probably due to the southward drainage of meltwater from the Eurasian ice sheet into the Aegean Sea from the Black Sea and Caspian Sea Basins (see Chapter 5). By contrast, drainage into the eastern Mediterranean from the Nile was greatly reduced due to increased aridity in eastern Africa.

Low latitudes and the tropics

The equatorward shift of polar high pressure areas and mid latitude cyclones resulted in a displacement and compression of the subtropical anticyclones between the mid latitude westerlies and the intertropical convergence zone (ITCZ). This led to the displacement of the trade wind belts (Figures 3.3 and 3.4). In western Africa, the southward shift of the NE trades resulted in a corresponding displacement of the summer SW monsoon that in turn caused monsoon precipitation over the Gulf of Guinea (Figure 3.3). In East Africa, a corresponding southward displacement of the SW monsoon by strengthened NE trade winds also caused monsoon precipitation over a much more restricted land area. By contrast, most areas of Africa during ice age winters appear to have been dominated by dry northerly and NE winds. As a result, the Kalahari Desert was subject to a major increase in its size during the last glacial maximum (Chapter 9).

A similar climatic regime may have occurred over northern areas of South America during the last glacial maximum (Figure 3.4). Here, displacement of the ITCZ over the area presently occupied by the equatorial rainforest may have caused a more restricted winter NE monsoon and pronounced aridity. By contrast, the southerly air streams that presently characterise western South America may have been more vigorous due to enhanced upwelling in the area of the Benguela Current along the eastern margin of a strengthened South Pacific anticyclone.

A result of these processes and also of the enhanced global meridional (north–south) flow in both hemispheres was the development over the major oceans of strengthened subtropical anticyclones due to more vigorous atmospheric circulation. This resulted in the development of very stable areas of subtropical high pressure and, by inference, the occurrence of relatively cloudless conditions. This may partly explain the CLIMAP reconstructions of ocean surface temperatures for 18,000 BP where large areas of the northern and southern Pacific Oceans as well as parts of the Indian Ocean are actually considered to have experienced surface water temperatures warmer (between +1 and +2°C) than present. Elsewhere over subtropical ocean areas, lower sea surface temperatures resulted in a contraction of those areas in which coral reefs developed (Goudie 1983).

Along the western margins of continents, enhanced oceanic upwelling appears to have resulted in the expansion of coastal deserts. Over adjacent ocean areas, more vigorous offshore winds led to increased aeolian deposition of terrigenous sediments on ocean surfaces. For example, large quantities of Saharan dust were deposited across the eastern Atlantic during the last glacial maximum (Sarnthein and Koopman 1980).

Tropical refugia and ice age aridity

Tropical refugia are isolated areas in which there has been sufficient continuity of favourable climate, soils, topography and vegetation to maintain the integrity of formerly more widespread landscapes and biotas (Street 1981). The small tropical rainforest refugia that existed during the last glaciation are today distinguished by a very large diversity in species of plants and animals, sometimes referred to as centres of endemism. For example, studies of the distribution of species of bird, woody angiosperms, butterflies and lizards in tropical South America have identified distinct refugia that were formerly separated by broad areas of semi-arid savanna (Bradley 1985).

In tropical Africa during the last glaciation there was also a marked reduction in the area of rainforest in the Congo and in Guinea (Hamilton 1976). Similarly, rainforest areas west of the Nile during the last glaciation were the exception rather than the rule with the principal refugia located in the highland areas of

Cameroon, eastern Zaire and Tanganyika–Malawi. The widespread tropical aridity implied by the presence of refugia is also illustrated by the occurrence of aeolian sediments in areas of present-day equatorial rainforest (Tricart 1974) and by the presence of aeolian sediments offshore (Damuth and Fairbridge 1970).

ICE AGE PALAEOCLIMATES IN THE SOUTHERN HEMISPHERE

Middle and high latitudes

In the southern hemisphere, the northward extension of sea ice cover to 50°S resulted in the northward displacement of the Antarctic polar atmospheric front. This resulted in a corresponding northward displacement of the mid latitude cyclones which moved over Argentina, southern Africa, southern Australia and Tasmania. The west to east movement of these cyclones and the patterns of snow precipitation are well illustrated by palaeoenvironmental reconstructions for the last glacial maximum of the Falkland Islands (52–53°S) in the South Atlantic and the Prince Edward Islands in the South Indian Ocean (45–47°S) (Figure 3.7). In the Falkland Islands there is no evidence for significant glaciation at 18,000 BP, a fact largely due to the position of the islands in the lee of the Patagonian ice sheet (see Chapter 4). In contrast, glaciation at 18,000 BP extended down to sea level over the Prince Edward Islands due to an abundant supply of snow precipitation from the northwardly displaced mid latitude cyclones that tracked eastward across the southern Indian Ocean.

The various reconstructions for the East Antarctic ice sheet for 18,000 BP (Denton and Hughes 1981) suggest that the ice sheet extended as far as the edge of the continental shelf (Figure 3.7) although this estimate is complicated by the effects of sea level lowering, grounding ice shelves and glacio-isostatic crustal deformation. In general, however, the ice sheet extended northwards by nearly 2° of latitude and was associated with a corresponding northward displacement of the Antarctic Divergence (Figure 3.7). However, the most fundamental change that affected atmospheric and oceanic circulation around Antarctica was the widespread development of sea ice which, during southern hemisphere winters, effectively doubled the area of surface ice cover and reached as far north as 55°S (Figure 3.7). Hays *et al.* (1976) estimated that during the last glacial maximum, the position of the Antarctic Polar Front was displaced north of its present position by as much as 7° of latitude. However, they also suggested that the position of the subtropical convergence was little changed. Thus periods of increased Antarctic glaciation were accompanied by a marked contraction in the area of sub-Antarctic surface waters.

The great increase in sea ice cover in the Antarctic Circumpolar Ocean during the last glacial maximum had a profound influence on the climate of the southern hemisphere and also may have increased the vigour of global oceanic and atmospheric circulation. The principal effect, caused by a greater sea ice cover, was an

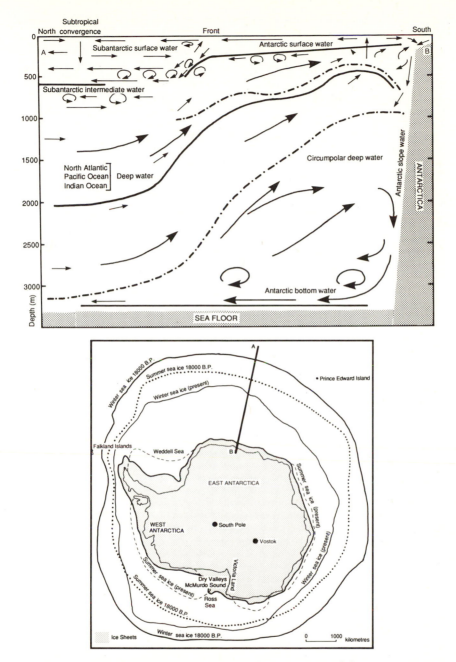

Figure 3.7 A. Vertical meridional circulation and water masses of the Antarctic Ocean (profile A–B) (after Lozano and Hays 1976). B. Winter and summer sea ice limits for the present and for 18,000 BP. Locations mentioned in the text are indicated as well as the location of profile line A–B (after various sources).

increase in size and strength of the area of Antarctic high pressure. This, in turn, may account for the northward displacement of the mid latitude cyclones as they tracked across the the southern Pacific, Indian and Atlantic Oceans. Elsewhere, a pronounced anticyclone developed over central Australia and led to the formation of extensive dune systems in areas located along the southern flanks of the anticyclone (see Chapter 9).

SUMMARY

The many studies of Late Quaternary palaeoenvironmental change that have been undertaken each provide information on past climatic conditions for particular areas of the world. Much of this information was gathered during the CLIMAP Project in an attempt to reconstruct the geography of the Earth during the culmination of the last glacial maximum at 18,000 BP. Despite the limitations and uncertainties associated with much of these data, the CLIMAP reconstructions provided, for the first time, a coherent geographical synthesis of ice age Earth.

Global climate at 18,000 BP was dominated by the great ice sheets that had developed in the northern hemisphere. Each ice sheet was associated with a permanent cell of high pressure. These anticyclones led to the southward displacement of mid latitude cyclones in the northern hemisphere and split jet stream flow in the upper troposphere. As a result, southwestern USA and the Mediterranean areas were subject to increased rainfall. Elsewhere, permanent high pressure over the Asian continent led to failure of the Asian monsoon and increased aridity over the Indian subcontinent. These changes were linked to changes in ocean circulation. In the North Atlantic, increased meltwater discharge led to more pronounced salinity stratification and a halt in the production of North Atlantic deep water.

The oceanographic changes were also associated with increases in the extent of sea ice cover in the North Atlantic, the Arctic Ocean and the Antarctic Circumpolar Ocean. In the North Atlantic the winter limit of permanent pack ice reached as far south as the latitude of Spain and New York while in the Antarctic Circumpolar Ocean, the winter sea cover reached as far north as 55°S. In both areas, the polar oceanic front was markedly displaced towards the equator. These changes also were accompanied by a corresponding displacement towards the equator of the polar atmospheric fronts. This led, in turn, to changes in the prevailing tracks of mid latitude cyclones in both latitudes.

Profound oceanographic and atmospheric changes also took place in low latitude areas. Strengthened high pressure cells over low latiude ocean areas led to the establishment over some ocean areas of sea surface temperatures that were actually higher than present. The overall trend, however, was of decreased temperatures over both low latitude oceans and continents. The reduction in the rate of North Atlantic deep water production appears also to have played a crucial role in ice age climate. Reduced NADW production also affected oceanic surface circulation in the Indian and Pacific Oceans. The effects of these, and other changes in the tropics, was to induce widespread aridity and the disappearance of any large areas of tropical rainforest that had existed previously.

The development of general circulation models began with the work of Gates who attempted to simulate global climate for 18,000 BP. Since then, there have been substantial improvements in the calibre of the GCMs. However, the model predictions are always dependent upon the boundary conditions that are used in the model. These can be subject to a great deal of uncertainty where they include various types of geological evidence. For example, disagreement about the sizes of the Laurentide and Eurasian ice sheets at 18,000 BP will lead to speculation about the accuracies of individual GCMs. This latter point illustrates quite clearly why global GCMs are dependent upon the results of empirical field and laboratory data, no matter how limited these are in their scope or content.

4

GLACIATION HISTORY FROM THE LAST INTERGLACIAL TO THE LAST GLACIAL MAXIMUM

INTRODUCTION

The extent and thickness of former glaciers and ice sheets has most frequently been determined by accurate field mapping of landforms of glacial erosion and deposition. Patterns of former ice flow direction are often reconstructed through measurements of the distribution of glacial erratics and by glacial striae orientations (Plate 1). The principal morphological features that have been used to define the positions of former ice margins are lateral and terminal moraines, fluvioglacial outwash plains, ice-marginal meltwater channels as well as the extent of drift sheets. Within the outermost limits of a particular glaciation, the presence of linear moraines may either indicate stages in the overall retreat and thinning of the ice or they may represent periods of glacier readvance. In many instances, individual landform assemblages are used to define the former extent of glaciers and ice sheets. It should be stated also that, in many instances, it is difficult to estimate the size of former ice sheets and valley glaciers since ice-marginal moraines may be absent. This can arise when the ice was relatively clean (and hence could not deposit any debris) or when the ice margin did not exist in a particular area for a sufficient length of time to allow the construction of moraines. The vertical extent of former ice sheets and glaciers is frequently aided by the distribution of periglacial landforms. In land areas located beyond or above the area directly affected by glacier ice, the boundaries between glaciated topography and the lower limit of frost- shattered bedrock and patterned ground phenomena, known as trim-lines, can be used to define the surfaces of former ice masses (Ballantyne 1990).

A major difficulty in calculating the dimensions of Late Quaternary ice sheets is that, because ice covered most land areas, there is no morphological evidence with which to define the position of former ice sheet surfaces. Thus, whereas there is abundant evidence to delimit the lateral extent of the ice sheets, there is very little information available to reconstruct former ice sheet profiles. Another difficulty arises due to the fact that most morphological evidence for earlier ice sheets has been removed owing to erosion by later ice advances. Not surprisingly therefore, there is abundant morphological evidence for the retreat of the last ice sheets since these have not been removed by glacial erosion. Evidence for the occurrence of earlier ice sheets is thus greatly dependent on stratigraphical information and accurate dating.

The interpretation of former glacier and ice sheet fluctuations is additionally complicated since end moraines can be produced under different environmental conditions. For example, recessional end moraines can be produced during periods of ice stillstand that interrupt the overall pattern of ice retreat. End moraines may also be formed during periods of ice readvance. However, glacial readvances need not always occur as a result of climate cooling. In many cases, particularly in circumstances where there there are excessive amounts of subglacial meltwater beneath the ice or where the ice terminates in water, glaciers may advance rapidly. These glacial surges are frequently unrelated to climate change. Therefore, in the reconstruction of past glacier and ice sheet fluctuations, it is of prime importance to differentiate those ice advances that took place due to surging from those attributable to climate change. In the following two chapters, these problems are highlighted by differences in interpretation of the patterns of glaciation and deglaciation for different parts of the world. In these chapters, an attempt is made to illustrate the regional complexity of Late Quaternary glaciation history and to describe the melting history of the last ice sheets.

A recurring theme in recently proposed models of global environmental change is the controversy over the sizes of the Late Quaternary ice sheets that formed, and the way in which they were produced. One view has been that the ice sheets which developed were largely terrestrially based and were relatively restricted in the scale of their ice cover. For example, Andrews (1973) suggested that the Laurentide and Eurasian ice sheets consisted of numerous land-based ice domes, the growth and decay of which were largely controlled by variations in precipitation. In this latter model, the Arctic Ocean had a sea ice cover during the last glacial maximum. A popular alternative view is that, in addition to landbased ice sheets, very extensive marine ice shelves and marine-based ice sheets were also formed. Attention has focused particularly on the likely influence of the Arctic Ocean on the growth and decay of the northern hemisphere ice sheets (Ewing and Donn 1956, 1958; Hughes *et al.* 1977; Denton and Hughes 1981), and one school of thought considers that very extensive marine-based ice sheets and ice shelves formed over the Arctic Ocean and Canadian Arctic archipelago during the last glacial maximum (Denton and Hughes 1981).

In the following discussion, reference is often made to the oxygen isotope stratigraphy for the Late Quaternary. In this respect, the ages for the isotope stages and substages are those described by Martinson *et al.* (1987) (Table 1.1). Reference to the Early, Middle and Late Quaternary time intervals is based on the criteria described in Chapter 1.

Plate 1 Sets of parallel friction cracks and glacial striae, Loch na Keal, Isle of Mull, Scottish Hebrides indicating a former pattern of ice sheet flow towards WSW (bottom right to top left). Photo S. Dawson.

EXTENT OF ICE COVER IN EURASIA AND THE ARCTIC

It has been suggested that the expansion of ice during the last (Late Valdai) glaciation, equivalent to oxygen isotope stage 2, was less extensive in the western USSR than its Early Valdai counterpart (Velichko 1984) (Table 4.1). Consideration of the available evidence lends support to the contention that during the Early Valdai (isotope stage 4), ice sheet expansion was more pronounced in Siberia and eastern Asia than it was in western USSR and Europe. By contrast, Late Valdai glaciation was more extensive in western USSR than in Siberia. The Middle Valdai period of interstadial warmth (isotope stage 3) appears to have been more pronounced in eastern Siberia than it was in the western USSR.

These suggestions are tentative since the development of the Eurasian ice sheet between 125,000 and 20,000 BP is not known in detail. Kind (1975) has suggested that a major period of ice sheet expansion in northeastern USSR may have taken place sometime prior to 40,000 BP. Kind (1975) also proposed that a later dramatic climatic deterioration took place in central Siberia between 33,000 and 30,000 BP (the Zhigansk stage). These two glacial periods are believed to have been separated in northeastern USSR by a period of warming (the so-called Karga interglacial) during which a major marine transgression took

Table 4.1 Late Quaternary chronostratigraphic divisions in Canada, USA and USSR

OXYGEN ISOTOPE STAGE	USA	CANADA		USSR	
1	Holocene		Holocene	European USSR	Siberia
10,000					
2	Late Wisconsin	Wisconsin Stage — Late	Late Valdai	Sartan
24,000					
3	(35)	Middle	Middle Valdai	Karga (non-glacial episode)
	Middle Wisconsin	Late Pleistocene			
59,000					
4	(65)				
74,000	Early Wisconsin	Early		
5a	(79)				
85,000		Sangamonian Stage	Early Valdai	Zyryanka
5b					
93,000					
5c	"Eowisconsin"			
105,000					
5d					
117,000					
5e	(122)				
125,000	Sangamonian		Interglacial	

After Velichko et al. (1984), Richmond and Fullerton (1986a), Martinson et al. (1987) and St Onge (1987). Not to scale.

Late Quaternary Chronostratigraphy in USA and Canada

A major difficulty in the Late Quaternary literature of Canada and the USA is the different chronostratigraphic terminology used in both countries. In Canada, the beginning of the Late Quaternary is placed on the basis of oxygen isotope stratigraphy at 130,000 BP at the beginning of the last (Sangamon) interglacial. However the boundary between the Sangamon and the Early Wisconsin is placed at 80,000 BP near the oxygen isotope stage 5/4 transition. Thereafter, the Early/Middle Wisconsin boundary is placed at isotopic stage boundary 4/3 until recently considered to occur at 65,000 BP, while the Middle/ Late Wisconsin boundary is placed at 23,000 BP. The Late Wisconsin/Holocene boundary is placed at 10,000 BP. It should be noted from the above that the chronostratigraphic terms 'Sangamonian' and 'Wisconsinan' are used respectively for the periods between 130,000 and 80,000 BP and between 80,000 and 10,000 BP (Fulton and Prest 1987).

The Late Quaternary chronostratigraphy used in the USA places the Sangamon/ Wisconsin boundary at the end of oxygen isotope substage 5e at 122,000 BP (Richmond and Fullerton 1986a). Thereafter, the Wisconsin glaciation is subdivided into four chronostratigraphic units. The first of these, the 'Eowisconsin', incorporates oxygen isotope substages 5a–d and ends at the isotope stage 5/4 transition considered to have occurred at 79,000 BP. In this way, the Early Wisconsin is believed to represent an isotope stage 4 between 79,000 and 65,000 BP while the Middle Wisconsin is considered (unlike Canada) to extend until 35,000 BP. The Late Wisconsin is therefore considered to occur between 35,000 and 10,000 BP.

45

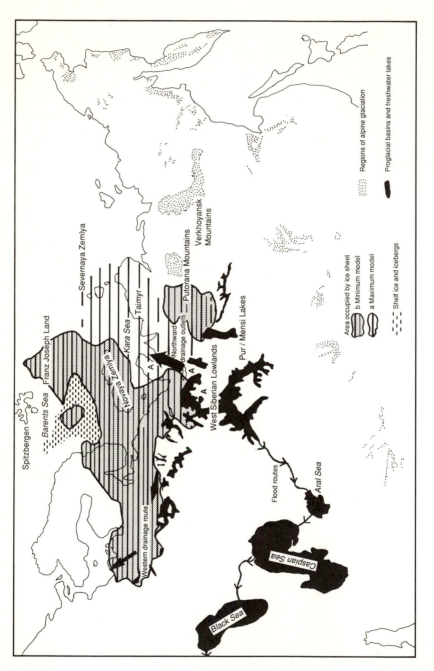

Figure 4.1 Late Valdai glaciation of the Soviet Union (after Velichko *et al.* 1984). The extent of ice in the West Siberian Lowlands (A) is uncertain.

place in Siberia. In central Siberia, the Zhigansk glaciation was very extensive. Both of these phases of ice sheet expansion were more widespread than any of the later phases of the Late Valdai (Sartan) glaciation during isotope stage 2.

The Karga period of relatively warm (interstadial) climate occurred between 50,000 and 25,000 BP when temperatures may have exceeded those of the present (Andersen 1981; Velichko 1984) while a similar period of warmth is recorded for European USSR at 50,000 BP (Velichko 1984). A later period of warmth is recognised for European USSR between 30,000 and 25,000 BP while in Japan a period of warmth is thought to have occurred prior to 34,000 BP with a secondary warm phase between 30,000 and 25,000 BP (Sakaguchi 1978). Sakaguchi (1978) also noted that in addition to the last glacial maximum at 21,000 BP, there was also a period of very cold climate at 31,000 BP.

The extent of ice in the USSR during the last (Sartan/Valdai) glacial maximum is also the subject of much disagreement. The principal controversy concerns the proposal that the last USSR ice sheet was coalescent with a large ice sheet and associated ice shelves in the Barents Sea, Kara Sea and Arctic Ocean and that the ice sheet also merged with the Fennoscandian ice sheet to the west (Figure 4.1). This view has been expressed by Hughes *et al.* (1977), Kvasov (1978), Grosswald (1980, 1984) and Denton and Hughes (1981). In contrast, numerous researchers (e.g. Velichko *et al.* 1984) have argued that the distribution of ice was considerably more restricted in these areas and that the Arctic Ocean possessed a sea ice rather than an ice shelf cover.

This difference of opinion is extremely important for several reasons. First, the disputed ice distributions have a considerable bearing on the estimation of the land ice/ocean water volume equivalents used to calculate the oxygen isotope parameters for the last glacial maximum. Second, the radically different ice distributions postulated for this region during the last glacial maximum are likely to produce quite different global atmospheric circulation patterns. Third, the pattern of Late Quaternary ice loading dictates the nature of glacio-isostatic land uplift and raised shoreline development throughout the Soviet Arctic. Finally, the Grosswald model of a very extensive Eurasian ice sheet implies the discharge of most proglacial meltwater from the southern margin of the ice sheet into the Mediterranean Sea via the Black Sea and Caspian Sea.

The Grosswald model of glaciation

Grosswald (1980) proposed that the ice cover during the Late Valdai/Sartan glaciation in the USSR formed part of a large Eurasian ice sheet that extended almost continuously from the west Siberian lowlands to Scandinavia and western Britain (Figure 4.1). He considered that, in the Soviet Arctic, the ice sheet formed separate domes over the shallow Barents Sea and Kara Sea continental shelf, much of which was exposed due to glacio-eustatic sea level lowering. These ice domes are thought to have acted as powerful centres of ice dispersal that resulted in a southward movement of ice from the Taimyr Peninsula into the North Siberian

Lowlands (Figure 4.1). Farther south, major centres of ice dispersal may also have occurred over the Ural Mountains and in the Putorana Mountains east of the Yenisei River. In eastern Siberia, a large independent ice cap existed in the Verkhoyansk Mountains (Figure 4.1).

During this period, the southern margin of the ice sheet extended from the Putorana Mountains westwards towards northern Poland, Germany and Denmark (Figure 4.1). Grosswald (1980) maintained that numerous large proglacial ice-dammed lakes were produced along the southern margin of the ice sheet that resulted in the submergence of 1.5 million km^2 of the West Siberian Lowland (Figure 4.1). In this model, large sections of the last Eurasian ice sheet terminated in water, a factor that may have significantly affected the development of ice-marginal moraines due to the influence of iceberg calving. The dating of the maximum phase of ice expansion has been derived largely from the radiocarbon dating of organic material buried beneath the outermost (Syrzi) moraines and of organic sediments buried beneath ice-dammed lake sediments. These suggest that the ice sheet advanced southwards shortly after 20,000 BP (Grosswald 1980).

The Velichko *et al.* model of glaciation

Velichko *et al.* (1984) disagree with the model of Late Valdai glaciation proposed by Grosswald. These workers have suggested instead that the presence of submerged moraines on the continental shelf of the Barents and Kara Seas as well as around Franz Joseph Land and Novaya Zemlya in the Arctic Ocean (Figure 4.1) suggests that ice accumulation was restricted to the larger islands and the high plateaux. Thus they consider that Late Valdai ice initially built up on the the Putorana, Ural and Taimyr Mountains as well as on the islands of Novaya Zemlya and Severnaya Zemlya. Thereafter, the ice sheet centre was displaced northwards onto the northern Taimyr Peninsula and the western part of the Kara Sea shelf (Velichko *et al.* 1984). In this model, the ice sheets that formed over Severnaya Zemlya and Taimyr were separated from the main ice sheet to the west by a large unglaciated area which approximately corresponds to the area between the rivers Ob and Yenisei (Figure 4.1). A consequence of this interpretation is that in the West Siberian Lowlands, the Late Valdai ice sheet may not have been sufficiently extensive to have enabled the formation of the large ice-dammed lakes described by Grosswald. As a result, Velichko *et al.*(1984) argue that much of the fluvioglacial discharge into the Caspian and Aral Seas proposed by Grosswald may not have taken place.

At present it is not possible to evaluate which of these two competing hypotheses of Eurasian glaciation is the more probable since much of the field mapping of the relevant ice-marginal moraines is incomplete. It is important, however, that the ages of the West Siberian moraines are accurately determined and that the principal episodes of ice-dammed lake development are more accurately dated.

Extent of glaciation in the Arctic

The controversy on the extent of the last glaciation in the Arctic ranges from the concept of complete ice cover (Denton and Hughes 1981) to partial glaciation (Boulton 1979). Dating of raised beaches and measurement of shoreline uplift isobases from Spitzbergen have been used by some authors (e.g. Grosswald 1980) to support the hypothesis of a complete Arctic Ocean ice shelf cover and by others (e.g. Boulton 1979) to demonstrate local glaciation of Spitzbergen during the last glacial maximum. A restricted ice cover for Greenland is suggested by Krinsley (1963) and Tedrow (1970) who have argued that parts of northern Greenland remained unglaciated during the last glacial maximum. Similarly, several areas of Arctic Canada may have remained free of ice during this period (Vincent 1978; Mayewski *et al.* 1981; Dyke 1983). The lack of glacier ice in these areas is considered to have resulted from precipitation starvation since a sea ice cover across the Arctic Ocean and North Atlantic, as well as the presence of the vast Laurentide and Eurasian ice sheets, led to the widespread development of permanent high pressure and hence arid conditions.

GLACIATION HISTORY OF SCANDINAVIA AND NORTHERN EUROPE

Early periods of ice growth and decay

Information on the rate and timing of glaciation in Scandinavia and northern Europe during the last (Weichselian/Devensian) glacial period is limited due to the erosion of pre-existing sediments by Late Weichselian ice. Some information is available, however, that points to the occurrence of several periods of glaciation since the last interglacial and prior to the Late Weichselian ice advance (Larsen and Sejrup 1990). The first of these is known in Scandinavia as the Gulstein Stadial and is considered to represent a period of ice expansion during oxygen isotope susbstage 5d (Figure 4.2). A second period of ice advance during the Eikelund Stadial is believed to have taken place during isotope substage 5b. The most extensive period of glaciation, however, is thought to have taken place during the transition between oxygen isotope stages 5 and 4 (Figure 4.2). This period of cold climate, known as the Karmøy Stadial, is thought to have lasted approximately 15,000 years and to have ended by 60,000 BP (Larsen *et al.* 1987; Larsen and Sejrup 1990).

In the North Sea region, Stoker *et al.* (1985) have described glaciomarine sediments of Early to Middle Devensian age from the central North Sea while Sutherland (1981) accounted for certain high-level marine deposits in Scotland as having been produced during a period of Early Devensian glacio-isostatic depression and ice expansion when world sea level was high. This period of ice growth may be equivalent in age to the Karmøy Stadial, having taken place during the oxygen isotope stage 5/4 transition considered by Ruddiman *et al.* (1980) to have been a period of extremely rapid ice growth in the northern hemisphere.

The inferred pattern of ice sheet fluctuations is in broad agreement with the climatic changes considered to have taken place elsewhere in Europe (Behre 1989). Thus two well-defined interstadial phases during isotope substages 5c (the Brorup interstadial) and 5a (the Odderdade interstadial) appear to have interrupted quite marked periods of cold climate during isotope substages 5d and 5b (Behre 1989) (Table 4.2).

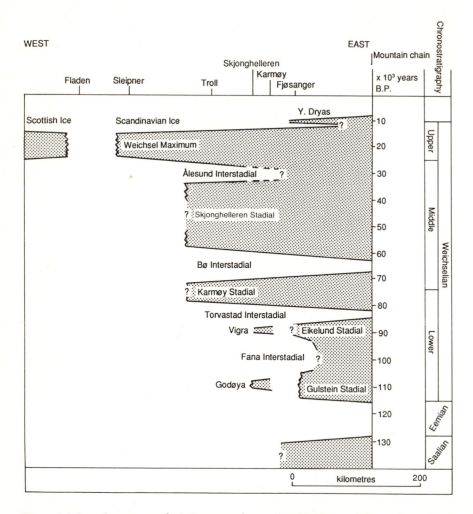

Figure 4.2 Late Quaternary glaciation curve for western Norway and the northern North Sea showing a complex history of ice sheet growth and decay (after Larsen and Sejrup 1990).

Table 4.2 Mean July temperatures in Netherlands, Grand Pile pollen record, chronostratigraphy for northern Europe, Norwegian glacial stratigraphy and oxygen isotope stratigraphy

WESTERN NORWAY ▥ Glacial Advance ■ Not Glaciated	NETHERLANDS Mean July Temperature (°C) 0 10 20	GRAND PILE ▥ Herbs ☐ Trees 50%	NORTH WEST EUROPE	Oxygen Isotope Stage	x10³ years B.P.
Holocene				1	10
Allerød Interstadial				2	24
Ålesund Interstadial			Denekamp	3	
			Hengelo/Upton Warren		59
			Oerel	4	74
Bø Interstadial		St. Germain II	Odderade	5a	85
			Rederstall	5b	93
Torvastad Interstadial		St. Germain I	Brørup S.S.	Brørup 5c	105
Fana Interstadial					
			Herning	5d	117
Fjøsangerian / Avaldsnes Interglacial		Eemian	Eemian	5e	

After Behre (1989) and Larsen and Sejrup (1990).

Later periods of ice growth and decay

Throughout NW Europe and Scandinavia, a period of climatic amelioration appears to have followed the cold conditions of isotope stage 4 (Lundqvist 1986a; Behre 1989) (Figure 4.2). An interstadial of this age in Sweden has been dated between 55,000 and 58,000 BP (the Jamtland interstadial) (Berglund and Lagerlund 1981; Lundqvist 1986a) and may equate to an interstadial of similar age in Finland (the Perapohjola interstadial) (Hirvas *et al.* 1981). More recently, Behre (1989) has provided evidence from Germany of a clearly delimited interstadial (the Oerel interstadial) that interrupted the cold conditions of oxygen isotope stage 4 (Table 4.2). Lundqvist (1986a) has argued that climatic amelioration during this period may have been accompanied by the virtual complete disappearance of Early Weichselian ice from the mountain regions of southern Norway, and also from most of Sweden. If correct, this is remarkable since the rapid northern hemisphere build-up of ice that commenced at 75,000 BP and is identified from ocean cores (Ruddiman *et al.* 1980) appears to have been followed by an equally rapid deglaciation that was completed by 60,000 BP. Additional support for this view is provided in the oxygen isotope record that shows a marked reduction in global ice volume at this time (Figure 2.2).

51

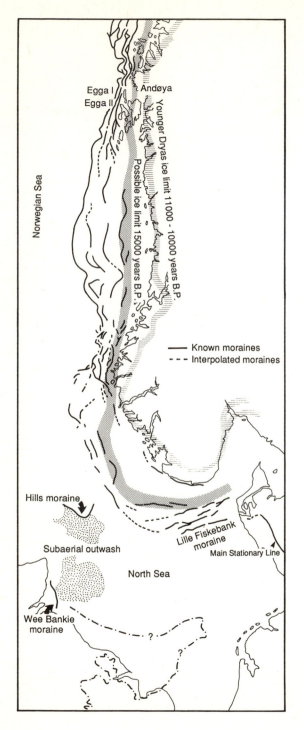

Figure 4.3 The distribution of glacier ice in the North Sea region during the last glacial maximum. The eastern limit of the British ice sheet is uncertain (dotted line). The distribution of ice-marginal moraines in Scandinavia is consistent with the view that much of the North Sea region was unglaciated during the last glacial maximum. The significance of the Hills moraine complex is not known (after Andersen 1979, Bowen 1978, Sutherland 1984; Long *et al.* 1988).

A later period of major ice accumulation is thought to have occurred in Scandinavia between 55,000 BP and 33,000 BP during the Skjonghelleren Stadial when the ice margin extended beyond the present coastline (Figure 4.2). The cold conditions may have been interrupted by a brief interstadial climatic amelioration during the Middle Weichselian/Devensian throughout NW Europe and Scandinavia. In Denmark, interstadial organic sediments have been dated to 37,000 BP (the Hengelo interstadial) (Houmark-Nielsen and Kolstrup 1981) while a similar brief period of warming appears to have occurred in Britain between 43,000 BP and 40,000 BP (the Upton Warren interstadial) (Mitchell *et al.* 1973; Shotton 1986) (Table 4.2).

The high relative sea levels of the Skjonghelleren Stadial

Additional evidence for the development of an extensive ice sheet over Scandinavia during the Skjonghelleren Stadial is provided by sea level data for the interstadial periods that preceded and followed this event. The interstadial prior to the Skjonghelleren Stadial (the Bø interstadial) (Figure 4.2) is considered to have been associated with a relatively low (10 m) relative sea level whereas the interstadial that succeeded the Skjonghelleren event (the Ålesund interstadial) between 33,000 and 28,000 BP occurred in conjunction with a relatively high (50–180 m) relative sea level. Although the extent of glacio-isostatic depression for this period is not known, one may speculate that the decline in relative sea level following the Skjonghelleren Stadial represents the influence of glacio-isostatic uplift following deglaciation. The extent of glacio-isostatic rebound in Scandinavia prior to Late Weichselian glaciation is also of particular importance since later ice sheet build-up during isotope stage 2 may have taken place upon a landmass that was not in glacio-isostatic equilibrium (see Chapter 11).

The last ice sheets in Scandinavia and northern Europe

The most significant expansion of ice throughout Scandinavia and northern Europe began sometime after 25,000 BP and culminated between 20,000 and 18,000 BP (Lundqvist 1986b). It has long been assumed that during this period (the Late Weichselian), the Scandinavian and British ice sheets were confluent (West 1972; Denton and Hughes 1981). However, there is an increasing body of information to suggest that this was not the case (Serjup *et al.* 1987) and that an ice-free corridor of dry land in the central North Sea separated the respective ice sheets due to sea level lowering (Figure 4.3) (Plate 2).

Much of the evidence for Late Weichselian end moraines off Norway is summarised by Andersen (1979) who described a remarkable sequence of large submerged moraines on the continental shelf west of Norway (the Egga I and Egga II moraines) (Figure 4.3). It is generally agreed that the Egga II moraines represent the limit of the Late Weichselian ice sheet. However, this scheme of Late Weichselian glaciation is inconsistent with the view that submerged end

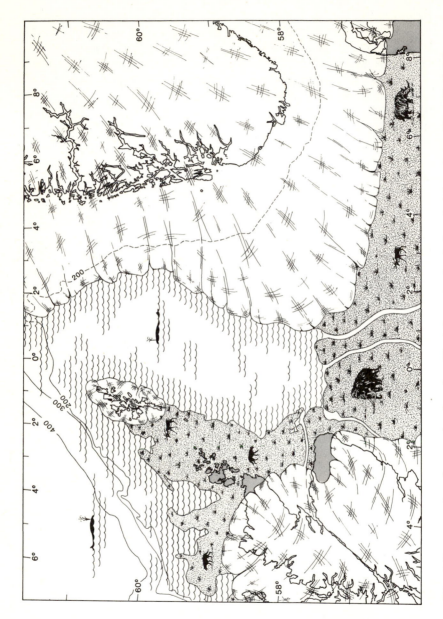

Plate 2 Reconstruction of geography of northern North Sea region for the last glacial maximum. Photo H.-P. Sejrup.

moraines and associated outwash (the Hills morainic complex) in the central North Sea are of the same age since they occur west of the Egga I moraines (Jansen 1976; Jansen *et al.* 1979; Sutherland 1984). Farther west, the eastern limit of the Late Devensian British ice sheet appears to be represented offshore by a north–south-trending feature interpreted as an end moraine in the Wee Bankie area (Holmes 1977; Sutherland 1984) (Figure 4.3). If this is the case, it is difficult to envisage the geometry of the eastern sector of the last British ice sheet since most authors consider that the ice sheet extended as far south as north Norfolk (Bowen 1978) (Figure 4.3).

Some authors (e.g. Jansen 1976) consider that the ice margin on the European continent extended across Jutland, Denmark and eastwards towards Frankfurt. The ice limit is also thought to extend westwards into the southern North Sea where it is represented by the Lille Fiskebank moraine (Andersen 1979) (Figure 4.3). However, this moraine is not dated and since it is located south of the Egga I moraine, it may possibly predate the last glacial maximum. In northern Germany, the Late Weichselian ice is considered to have deposited the extensive Brandenburg moraines that extend eastwards into Poland and are continued farther east across western USSR (Denton and Hughes 1981; Grosswald 1980) (Figure 5.1).

In northern Norway, several islands (e.g. Andøya) appear to have remained ice free during the Late Weichselian (Vorren *et al.* 1988) (Figure 4.3). This view is consistent with the view that, at the time of the Late Weichselian ice maximum, the polar oceanic and atmospheric fronts in the North Atlantic were displaced southwards as far as the latitudes of southern Spain and New York. Precipitation starvation, caused by the southward migration of the polar fronts, may also be invoked to account for the occurrence of unglaciated areas in northern Scotland during the last glacial maximum (Sutherland 1984).

LATE QUATERNARY ICE SHEETS IN NORTH AMERICA

Introduction

In any consideration of global environmental changes during the Late Quaternary, the growth and decay phases of the Laurentide and Cordilleran ice sheets are of fundamental importance. The Laurentide ice sheet during the Late Wisconsin had a volume of between 18 and 35 million km^3 of ice and extended across nearly 4,400 km of the North American continent (Fulton and Prest 1987) (Figure 4.4). To the west, it merged with the Cordilleran ice sheet that, in turn, extended 3,700 km westwards from the Puget Sound Lowland, British Columbia, across southern Alaska to the Aleutian Islands (Figure 4.5). As a result, most of the northern half of North America was buried by an ice mass that constituted more than one-third of the world's glacial cover (Flint 1971). The ice sheets also, by virtue of their great thickness, exerted a profound topographic control on ice age atmospheric circulation (see Chapter 3).

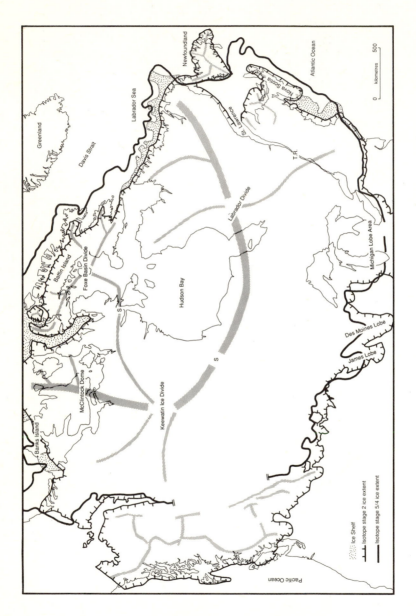

Figure 4.4 The possible extent of the Laurentide ice sheet during oxygen isotope stage 2 (hachured thin line) and during the stage 5/4 transition (thick line) (after Vincent and Prest 1987). The positions of major ice divides (Andrews *et al.* 1986) and principal locations mentioned in text are also shown. T.R. = Trois Rivières.

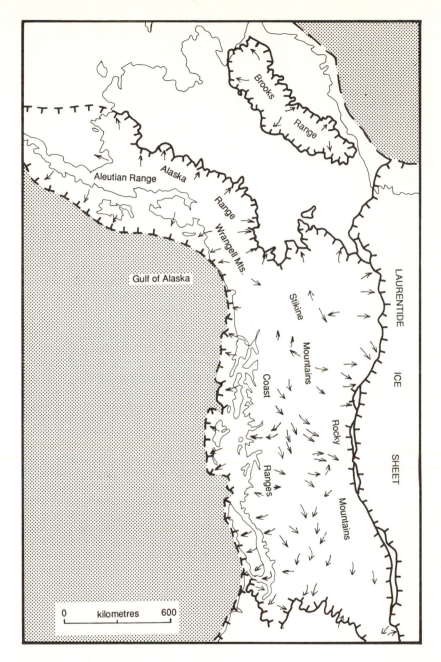

Figure 4.5 Generalised map showing approximate extent of Cordilleran ice sheet. Arrows indicate former ice flow directions. Note the inferred presence of extensive north Pacific ice shelves and the aridity that prevailed over much of central and northern Alaska (after Flint 1971).

57

The Late Quaternary evolution of the Laurentide and Cordilleran ice sheets appears to have been characterised by non-synchronous advances and retreats of ice along different sections of the ice sheet margin. Furthermore, the growth of the southern sectors of the Laurentide ice sheet may have been accompanied along the northern margins by precipitation starvation and ice retreat (Fulton 1989). Andrews (1982) has argued strongly that the Laurentide ice sheet was not stable cored but rather a complex and dynamic system with at least three or more centres of growth that did not respond to climatic change in a synchronous manner. He also pointed out a serious anomaly in that available reconstructions of the Laurentide ice sheet depict ice divides that are inconsistent with the locations of maximum ice thickness as inferred from the measured patterns of glacio-isostatic uplift.

Early periods of ice advance

In the following discussion, usage of the term 'Eowisconsin' broadly refers to isotope substages 5a–d while the term 'Early Wisconsin' approximately equates to oxygen isotope stage 4 (Table 4.1). The field evidence in North America for 'Eowisconsin'(i.e. 122,000 to 79,000 BP) glacier fluctuations is partly derived from the Canadian Arctic and also from the Yellowstone area, Wyoming, where an independent ice cap developed separately from the large ice sheet to the north. In Yellowstone, Richmond (1986b) concluded that two glacial advances took place during isotope substages 5d and 5b. Glacial deposits of similar ages have been identified from Banks Island and eastern Baffin Island in Arctic Canada and also from the Hudson Bay Lowlands (Andrews *et al.* 1983; Shilts 1984) (Figure 4.6). For example, Andrews *et al.* (1983) have suggested that a major ice advance took place from Labrador into the Hudson Bay Lowlands during isotope substage 5d. According to Karrow (1984), evidence for climatic cooling and the possible presence of ice during oxygen isotope substages 5b–d in the St Lawrence valley is indicated by the deltaic sediments of the Scarborough formation near Toronto and the Becancour till in the central St Lawrence Lowland.

During this period, ice advance may also have occurred over eastern Baffin Island (the Ayr Lake Stadial) from the Foxe Basin while ice may also have advanced over Banks Island from Keewatin (the M'Clure stadial) (Andrews and Miller 1984; Shilts 1984; Vincent 1984; Andrews *et al.* 1986) (Figure 4.6). The latter authors have argued on the basis of raised shoreline evidence that ice accumulation during this period was associated with considerable glacio-isostatic depression.

A contrary view has been expressed by Denton and Hughes (1981: 440) who have suggested that because the oxygen isotope record indicates a fall of global sea level of greater than 75 m between 124,000 and 115,000 BP, extensive ice shelves are likely to have become grounded in Hudson Bay, Foxe Basin and elsewhere throughout the Canadian Arctic archipelago. Thus Denton and Hughes envisaged that ice sheets may have developed from grounded ice shelves in these areas. By contrast, Andrews and Mahaffy (1976) argued that the raised shoreline evidence from the Canadian Arctic demonstrates quite convincingly that most ice accumulation was land based and occurred in conjunction with several largely independent ice masses.

The scale of Early Wisconsin glaciation (between 79,000 and 65,000 BP) over North America is highly controversial and is well reviewed by Vincent and Prest (1987). Some authors (e.g. Dredge and Thorleifson 1987) have considered that large areas of ice that formed during isotope stage 4 may simply represent ice which developed initially during isotope substages 5a–d and which persisted throughout the intervening period. The Early Wisconsin glaciation most probably began during the transition between isotope stages 5 and 4 at 75,000 BP.

Richmond and Fullerton (1986a: 191) have suggested that glaciation during stage 4 was particularly extensive throughout east central and eastern North America but was not extensive or did not occur within the Michigan lobe area of Illinois and southern Michigan and from southern Wisconsin westwards to the Rocky Mountains in Montana. There was also considerable development of glacier ice in the Coast and Cascade Ranges of NW USA and western Canada

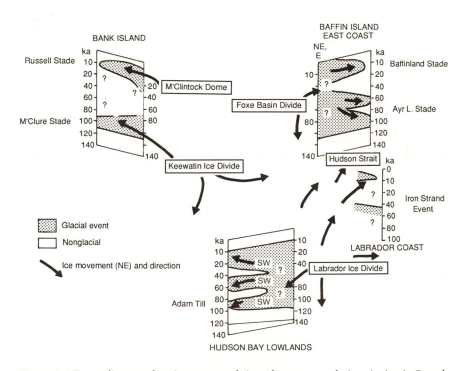

Figure 4.6 Fence diagram showing suggested time/distance correlations in Arctic Canada and possible correlations with southern records (after Andrews *et al.* 1986). The locations of the principal ice divides are shown on Figure 4.4.

(Easterbrook 1976; Armstrong and Clague 1977) but it is not known whether these ice masses formed part of a Cordilleran ice sheet or if they reflect alpine valley glaciation during this period. In the 'Ice-Free Corridor area' of eastern British Columbia and western Alberta the extent of ice accumulation at this time is unknown although it is generally accepted that ice expansion during this period was more extensive than during the Late Wisconsin (Clayton and Moran 1982; Fulton *et al.* 1986a). In the Canadian Arctic, ice appears to have advanced over eastern Baffin Island between 80,000 and 40,000 BP (Andrews *et al.* 1983; Andrews *et al.* 1986: 247) while a similar ice advance may have occurred over the Hudson Bay Lowlands at this time (Figure 4.6). A similar phase of Early Wisconsin glaciation is recognised for Alaska (Hamilton 1986a,b).

The possibility that glaciation in western North America was restricted during this period and that considerable ice sheet growth was correspondingly taking place in eastern North America may be indicative of the changeover of North Atlantic oceanic and atmospheric circulation to a glacial mode and the supply of snow precipitation from a neighbouring warm North Atlantic Ocean. It may also have been related to jet stream convergence above the eastern margins of the Laurentide ice sheet. Vincent and Prest (1987) have suggested that glacial build-up during this period may have been considerably greater than during the Late Wisconsin glaciation, not only throughout the Canadian Arctic but also throughout the Atlantic Appalachian region (Figure 4.4). This considerable ice sheet growth that took place at this time may coincide with a period of relatively warm ocean temperatures that resulted in an enhanced supply of snow precipitation (Ruddiman *et al.* 1980).

Non-glacial intervals

A limited amount of information is available on non-glacial deposition in North America during oxygen isotope substage 5a. For example, St Onge (1987) has argued that the St Pierre organic sediments near Trois Rivières in the central St Lawrence Lowland may indicate interstadial sedimentation at 75,000 BP. Similarly, Andrews *et al.* (1983) concluded that fluvial deposition of the Fawn River gravels in the Hudson Bay Lowlands took place during an interstadial at 76,000 BP (Figure 4.6). The palaeocurrent indicators from these gravels, showing discharge to the north, led Andrews *et al.* (1983) to conclude that Hudson Bay and Hudson Strait were deglaciated (as presumably were other centres of ice dispersal) by the end of substage 5a.

A Middle Wisconsin episode of general climatic amelioration is widely recognised in North America but the age of its boundaries varies according to the terminology used (Dreimanis and Raukas 1975; Richmond and Fullerton 1986a,b) (Table 4.1) although the period broadly corresponds to oxygen isotope stage 3. This isotopic stage is generally regarded as a time of climatic amelioration and retreat and decay of Laurentide and Cordilleran ice (Sancetta *et al.* 1973). Large-scale melting of the Cordilleran ice sheet may also have taken place during the Middle Wisconsin and it has been suggested by Clague (1981) and Fulton *et al.* (1986a) that the ice sheet may have melted completely during this period. Thus, Gascoyne *et al.* (1980) have suggested from speleothem studies that large areas of western Canada were not glaciated between 67,000 and 28,000 BP, while in Alaska the truncation of ice wedges during moist conditions between 36,000 and 30,000 BP appears to have been followed by loess deposition and glacial advance after 30,000 BP (Pewe 1975; Hamilton 1986a,b).

The scale of continental deglaciation during isotope stage 3 is the subject of considerable disagreement while the volumes of ice that remained on the continental shelf and in the continental interior are largely unknown. Three periods of interstadial warmth characterised by cool-climate forest-tundra vegetation have been described by Dreimanis *et al.* (1966) for Port Talbot, Lake Erie. The first of these (Port Talbot interstadial I) commenced before 48,000 BP while the second (Port Talbot interstadial II) appears to have culminated between 40,000 and 47,000 BP. The third interstadial (the Plum Point interstadial) appears to have occurred between 25,000 and 30,000 BP (Dreimanis *et al.* 1966; Dredge and Thorleifson 1987). These interstadials may have been interrupted by major phases of ice expansion in the Lake Ontario Basin (Fulton *et al.* 1986b) although a limited (or no) ice cover is suggested for many areas.

Non-synchronous changes across the northern hemisphere?

It appears that the European interstadial events at 59,000–60,000 BP have no age counterpart in North America. Nevertheless, despite numerous dating uncertainties, one or two periods of interstadial warmth appear to have taken place in North America during isotope stage 3. However, correlation of these events with the oxygen isotope record and the inferred global ice volume changes remains uncertain.

The pattern of Middle Wisconsin glaciation and deglaciation history in North America may be contrasted with the chronology proposed for the USSR where regional deglaciation is believed to have been interrupted by major ice sheet expansion between 42,000 and 35,000 BP and from Scandinavia where major ice accumulation took place during the Skjonghelleren Stadial between 55,000 and 33,000 BP. However, Alam and Piper (1977) have provided ocean floor microfaunal data from offshore eastern Canada that demonstrate drastic cooling between 40,000 and 30,000 BP with conditions even more severe than during the Late Wisconsin. The global significance of this cooling phase remains unclear although it hints at the possibility that very severe glacial conditions prevailed in the northern hemisphere between 40,000 BP and 35,000 BP.

The Late Wisconsin glaciation of North America

The Late Wisconsin glacial stage in North America is generally considered to extend between 35,000 and 10,000 BP (Table 4.1). During this period the Laur-

61

entide ice sheet was confluent with Cordilleran ice which together covered an area of 16 million km² and constituted, together with Greenland, almost 40 per cent of the global ice cover during that period (Flint 1971). In NW Canada, the large Cordilleran ice mass developed between the Coast Ranges, the Cascades and the Rocky Mountains to the east (Figure 4.5). The predominant source of snow precipitation was the Gulf of Alaska and, as a combined result of topographic influence and precipitation distribution, the Pacific coastline of British Columbia and Alaska was dominated by the presence of tidewater outlet glaciers and floating ice shelves. Hence, most precipitation derived from westerlies over the NE Pacific resulted in ice accumulation in the area west of the Rocky Mountains.

By contrast, the area east of the Rockies was effectively deprived of precipitation owing to the presence of adiabatically warmed air in the lee of the Rockies (see Flint 1971). The Laurentide ice sheet during the Late Wisconsin extended from the eastern foothills of the Rockies to the NW Atlantic Ocean east of Nova Scotia and Newfoundland. During this period, large areas of the Cordilleran and Laurentide ice sheets were confluent although the timing of ice advance is not known with certainty (Clayton and Moran 1982; Fulton *et al.* 1986a; Vincent and Klassen 1989). North of 60°N, the Laurentide ice merged with ice over Ellesmere Island and Baffin Island and may have been confluent with the northwest section of an expanded Greenland ice sheet.

Several coalescing ice caps developed over the Aleutian Islands during the Late Wisconsin, with the former ice divide being located south of the archipelago (Figure 4.5). This resulted in a northward flow of ice across the islands where it terminated in the southern Bering Sea as a floating ice shelf (Thorson and Hamilton 1986). In Alaska, north of the Cordilleran ice sheet, a local ice cap also developed over the Brooks Range (Pewe 1975; Hamilton 1986a,b) (Figure 4.5). However, central and northern Alaska were characterised by aridity, a factor largely due to the isolation of the area by the Cordilleran ice from the prime source of moisture in the Gulf of Alaska. Pewe (1975: 108), however, has argued that the scale of Wisconsin glaciation in Alaska may be attributable to a decrease in mean summer temperatures (of 3–4°C) rather than to any changes in the precipitation regime.

The build-up of the Laurentide ice sheet during the Late Wisconsin appears to have commenced about 25,000 BP. In western Canada, the Late Wisconsin Cordilleran glaciation may have commenced upon an ice-free land surface. However, it is suspected that large areas of eastern Canada remained ice covered throughout the Middle Wisconsin and therefore that the build-up of the Laurentide ice sheet during the Late Wisconsin was associated with the growth and expansion of pre-existing ice masses.

Several authors have stressed that the maximum of Late Wisconsin glaciation did not occur at 18,000 BP as assumed by CLIMAP reconstructions (cf. Fullerton and Richmond 1986). Instead, numerous radiocarbon dates suggest that the Cordilleran glacial maximum occurred nearer 15,000 BP while the maximum accumulation of the Late Wisconsin Laurentide ice sheet appears to have been between 20,000 and 21,000 BP. This in turn suggests that most Late Wisconsin ice sheet growth took place during a circa 5,000 year interval between 25,000 and 20,000 BP. Along the southern margin of the Laurentide ice sheet, numerous topographically controlled ice lobes resulted in the development of extensive end moraine systems. However, major lobes advanced/surged in the Great Lakes region for several thousand years while farther west the Des Moines Lobe extended south to central Iowa at 14,000 years BP (Figure 5.6A).

In general, the Late Wisconsin Laurentide ice sheet in eastern North America was not everywhere as extensive as the ice sheets that advanced between 122,000 BP and 65,000 BP although the extent was similar in many areas (Figure 4.4). Similarly, the sequence of Alaskan glaciation during the Wisconsin indicates that the Late Wisconsin ice advances were more restricted than those that took place earlier during the Late Quaternary. These older ice sheets appear to have been slightly larger in size with probably a more active centre of ice dispersal over Labrador and considerable ice advance over the Scotian Shelf (Vincent and Prest 1987) (Figure 4.4). The more restricted size of the Late Wisconsin Laurentide ice sheet in eastern North America may have been due to the considerably reduced North Atlantic Ocean temperatures during isotope stage 2 which led to a decreased supply of snow precipitation.

REST OF THE WORLD

Introduction

The chronology of Late Quaternary glacier and ice sheet fluctuations in other areas of the world is subject to much uncertainty. For example, there is a great deal of disagreement about the scale of Late Quaternary glaciation in China and Tibet, a factor of crucial importance to our understanding of Chinese loess development as well as to the influence of Tibetan uplift on global climate change. The Late Quaternary glacial history of Antarctica is also poorly understood although the subject is of prime importance to our understanding of the dynamics of global climate change. The chronology of Late Quaternary mountain glaciations is also not well known yet is of vital significance since it may tell us if glaciations in high and low latitudes of both hemispheres took place synchronously. In the following sections these issues are considered briefly: they serve to demonstrate how much we still do not know about past glacier fluctuations.

Late Quaternary glaciation of China

Much of the pioneering research on the Quaternary glaciation of China was undertaken by Li Siguang who established a pattern of glaciation and interglaciation based on the model of alpine glaciation proposed by Penck and Bruckner (1909). The most recent research, however, suggests a very complex history of glaciation that is still poorly understood (Shi *et al.* 1986). Most present glaciers in China occur in the Qinghai-Xizang Plateau (Tibet) with a smaller number of glaciers developed in the Kunlun Shan Range. It is argued by many workers that two major phases of glaciation took place in these regions during the Late Quaternary, possibly during oxygen isotope stages 2 and 4 (Shi *et al.* 1986). The latter authors have stressed, however, the great importance of Late Quaternary neotectonic uplift of the Qinghai-Xizang Plateau as a factor affecting the glaciation of China. For example, it is estimated that an astonishing 3,000 m of uplift has taken place on the Tibetan Plateau during the Quaternary and that a considerable proportion of this took place during the Late Quaternary (see Chapter 11) (Tungsheng *et al.* 1986; Shi *et al.* 1986). This, in turn, is likely to have resulted in a reduction in snow precipitation, the weakening of monsoonal circulation and the development of progressively smaller Tibetan glaciers. Shi *et al.* (1986) and Benxing (1989) are emphatic in their view that there was no continental ice sheet in western China during the Late Quaternary. A contrary view has been expressed by Kuhle (1987, 1988) who has proposed that, during the Quaternary, Tibet was covered by a continental ice sheet with an area of 2–2.4 million km^2. According to Kuhle (1987, 1988), the presence of such a vast ice sheet during the Late Quaternary, accompanied by an equilibrium line altitude (ELA) lowering of 1,100–1,600 m, would have played a critical role in the development of global atmospheric circulation during glacials. The resolution of this difference of opinion is vital if more accurate global climate models are to be produced (Benxing 1989). It also will be of great value in understanding the origin of Chinese loess since, if it existed, a large Tibetan ice sheet would have provided an abundant source of 'cold' loess (see Chapter 9).

Late Quaternary glaciation history of Antarctica

The present volume of Antarctic ice is 24 million km^3 of which almost 83 per cent constitutes the East Antarctic ice sheet (Table 4.3). The East Antarctic ice sheet presently reaches a maximum surface altitude of 4,000 m and is developed on a land surface most of which occurs above sea level. In contrast, most bedrock beneath the West Antarctic ice sheet occurs below sea level while the ice sheet itself is flanked by extensive floating ice shelves.

The glaciation history of Antarctica between oxygen isotope stages 5e and 2 is at present largely unknown. Most information is available regarding the extent of the Antarctic ice sheet during the last glacial maximum. However, even here, there is controversy. One view is that both the West Antarctic ice sheet and the East Antarctic ice sheet extended near the continental shelf and reached their maximum positions between 17,000 and 21,000 BP and probably nearer 20,000 BP (Heusser 1989). By contrast, it has also been proposed that the West Antarctic ice sheet was not significantly larger during the last glacial maximum than it is today (for a fuller discussion see Denton and Hughes 1981: 290). The latter authors concluded that the total ice volume of both the East and West

Table 4.3 Relationship between volumes of last ice sheets and global sea level lowering according to the minimum and maximum ice reconstructions of Denton and Hughes (1980). The values assume isostatic equilibrium and a rock–ice ratio of 4

	Minimum Reconstruction		Maximum Reconstruction	
Ice volume	Total volume (10^6 km^3)	Volume causing lower sea level (10^6 km^3)	Total volume (10^6 km^3)	Volume causing lower sea level (10^6 km^3)
Ice sheets				
Laurentide	30.900	30.500	34.800	34.200
Cordilleran	0.260	0.260	1.900	1.840
Innuitian			1.130	0.983
Greenland[a,b]	2.920[a]	0.287[b]	5.590[a]	2.550[b]
Greenland[b,c]	0.287[c]	0.287[b]	2.950[c]	2.550[b]
Iceland	0.050	0.050	0.267	0.236
British Isles	0.801	0.773	0.801	0.773
Scandinavian	7.250	7.060	7.520	7.320
Barents–Kara	0.955	0.865	6.790	6.250
Putorana			0.581	0.581
Antarctica	37.700	9.810	37.700	9.810
East	24.200	3.330	24.200	3.330
West	13.500	6.480	13.500	6.480
Glaciers and ice caps	1.184	1.830	0.750	0.715
Totals	84.174	51.300	97.829	65.300
Sea level equivalent[d]		127 m		163 m
Eustatic sea level drop[e]		91 m		117 m

After Denton and Hughes (1981)

[a] Total Late Wisconsin–Weichselian ice.
[b] Additional Late Wisconsin–Weichselian ice contributing to lower Late Wisconsin–Weichselian sea level.
[c] Additional Late Wisconsin–Weichselian ice.
[d] Without hydro-isostatic sea floor rise, using present ocean area (361×10^6 km^2).
[e] With hydro-isostatic sea floor rise.

Antarctic ice sheets during the last glacial maximum was 37 million km^3 due mostly to an increase in the size of the East Antarctic ice sheet. A result of this ice sheet expansion was a 25 m lowering of global sea level.

During the last glacial maximum, the majority of the ice expansion took place over West Antarctica. It has been argued that the expansion of the West Antarctic ice sheet, in conjunction with sea level lowering during the last glacial maximum, resulted in the grounding of ice on the Ross Sea and Weddell Sea continental shelves. By contrast, the expansion of the predominantly land-based

East Antarctic ice sheet was slight and perhaps, due to the narrow surrounding continental shelf, only as far as 75 to 90 km beyond its present-day margin (Hollin 1962; Denton and Hughes 1981). The prevailing view, however, of Antarctic ice sheet fluctuations during the Late Quaternary is that they were primarily influenced by global sea level changes which, in turn, were largely influenced by the growth and decay history of the Laurentide and Eurasian ice sheets. Thus, during periods of major ice sheet expansion in North America and Eurasia, the lowered global sea level resulted in the grounding of West Antarctic ice on both the Ross Sea and Weddell Sea shelves and the consequent expansion of the West Antarctic ice sheet. This model suggests, therefore, that Antarctic ice sheet fluctuations were probably approximately in phase with those of the northern hemisphere ice sheets (Hollin 1962; Denton and Hughes 1981; Labeyrie *et al.* 1986; Heusser 1989).

The development of the West Antarctic ice sheet during the last glacial maximum is, however, complicated, by the possibility that areas of the ice sheet may have retreated during this period due to decreased precipitation supply caused by the expansion of sea ice (Mercer 1983). Stuiver *et al.* (1981) and Mercer (1983) have, for example, demonstrated the retreat of alpine glaciers over south Victoria Land during the last glacial maximum due to precipitation starvation. A more complex scenario has been proposed by Denton *et al.* (1971) for the Antarctic Dry Valleys of south Victoria Land. These authors considerd that during periods of global glaciation when ice was grounded in the McMurdo Sound and the Ross Sea, ice advanced into the Dry Valleys at a time when nearby land-based glaciers were retreating due to decreased precipitation (Figure 3.7).

There is very little information available on the nature of Antarctic ice sheet fluctuations between the last interglacial and the last glacial maximum. Omoto (1977) considers that the East Antarctic ice sheet was characterised by ice retreat prior to 30,000 BP. However, there is insufficient available data to permit even the most rudimentary reconstructions of ice dimensions. Of considerable significance, however, is the suggestion that the West Antarctic ice sheet disintegrated during the last interglacial at 125,000 BP (Denton and Hughes 1981; Mercer 1978). These authors have argued that an increase in sea surface temperatures in conjunction with relatively high interglacial sea levels (i.e. similar to present) in West Antarctica would have resulted in the rapid disintegration of the Ross and Weddell Sea ice shelves and that once this had taken place, the inland ice domes would also have disintegrated since they were no longer buttressed by ice shelves. Although there is no geological evidence to indicate that such a major ice disintegration took place, some would argue that such widespread ice melting is implicit in the observation of exceptionally high global sea levels during oxygen isotope substage 5e.

An intriguing hypothesis was suggested by Wilson (1964) who argued that major surging of parts of the West Antarctic ice sheet may have taken place near the end of isotope substage 5e. Wilson (1964) proposed that the rapid influx of fresh water in the world's oceans together with an increase in albedo may have triggered widespread glaciation. Such a dramatic surge is likely to have raised global sea level – perhaps by as much as +17m (Hollin 1969). However, although there is evidence that considerable ice sheet growth took place during isotope substage 5d, there is at present insufficient evidence to demonstrate that high sea levels occurred worldwide near the end of substage 5e as a result of such an ice surge.

Mountain glaciers and small ice caps

The culmination of the last glacial maximum is recorded by a 40,000 km^2 ice cap over South Island, New Zealand (Porter 1975) that may have attained its maximum development at 19,000 BP (Suggate and Moar 1970). Elsewhere, widespread ice accumulation took place in the South American Andes, particularly in Patagonia between Peru and southern Chile (Caldenius 1932; Paskoff 1977; Clapperton 1983) (Figure 4.7). The Patagonian ice cap occupied an area of 480,000 km^2 and extended over 2,000 km in length and was up to 400 km wide (Clapperton 1983). The ice cap was associated with snowline lowering of up to 1,000 m (Denton and Hughes 1981; Clapperton 1983). The glacial history of both areas is complicated by Late Quaternary tectonic uplift that in some areas has been considerable. For example, in South Island, New Zealand, convergence of the Pacific and Australian plates has resulted in average uplift rates since the last interglacial of between 3.2 and 7.8 m per thousand years (Bull and Cooper 1986). Similarly, the Quito area of Ecuador is believed to have been uplifted by circa 150 m during the last 50,000 years (Clapperton 1987: 859).

In the Chilean Lake District, the maximum of the last glaciation occurred between 30,000 and 19,000 BP (Porter 1981; Mercer 1983). This period of glacier expansion appears to have been preceded by a more extensive earlier phase that tentatively may have occurred during oxygen isotope stage 4 (Mercer 1983). Van der Hammen et al. (1981) concluded that ice expansion during the last glacial maximum was less than during preceding glaciations, a pattern that is consistent with that observed for eastern North America (Vincent and Prest 1987). An early glacial phase is also recognised for the Ecuadorian Andes by Clapperton (1987) who suggested that a very extensive glaciation took place here between 36,000 and 45,000 BP (see also Van der Hammen et al. 1981).

In Europe, Herail et al. (1986) have noted the occurrence of a period of Weichselian glaciation in the French Pyrenees that took place prior to 40,000 BP (stage 4?) that was more extensive than a later (20,000 BP) glaciation. In the French and Italian alpine piedmont, Billard and Orombelli (1986) have noted that interstadial deposition (the oldest dated to 64,000 BP and the youngest to 29,000 BP) was followed by major ice advance that culminated at 18,000 BP. The limit of the last glacial maximum is well defined for Switzerland and Austria but the extent of Early Weichselian ice is not known with any certainty.

In numerous other areas, the culmination of the last glacial maximum was characterised by small areas of ice accumulation. For example, relatively small-

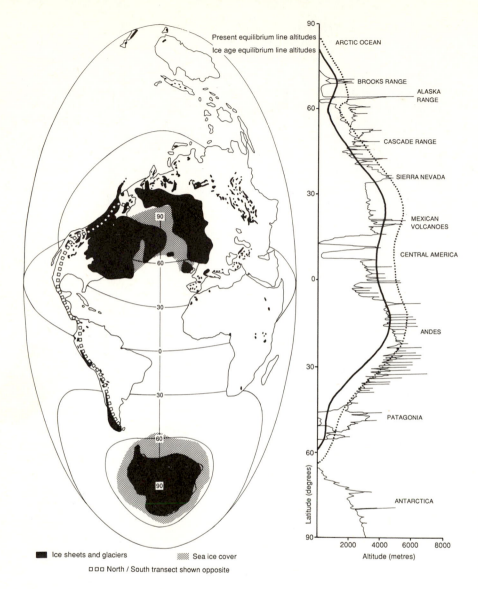

Figure 4.7 The distribution of the principal ice sheets and mountain glacier complexes during the last glacial maximum (left). The worldwide lowering of equilibrium line altitudes during the last glacial maximum is depicted along a north–south transect (open boxed line) (after Broecker and Denton 1990a).

scale glacier development took place in Tasmania and New South Wales, Australia, where the maximum ice expansion occurred at 20,000 BP (Bowler *et al.* 1976). Porter (1979) concluded that two major periods of Late Quaternary glaciation occurred on Mauna Kea, Hawaii. He noted that the first of these (the Early Makanaka glaciation) may have occurred during oxygen isotope stage 4 while the second and later glaciation (Late Makanaka) reached its maximum size sometime after 30,000 BP.

In several areas of the tropics, the glacial maximum seems to have occurred relatively late. For example, Mercer (1983) concluded that it may have culminated in some areas of tropical South America as late as 14,000 BP. Clapperton (1983) observed that throughout the Andes a significant ice advance occurred between 16,000 and 14,500 BP. Similarly, Livingstone (1962) suggested that the last glacial maximum in the Ruwenzori, west of Mt Kenya, may have occurred shortly before 15,000 BP. Also glaciers appear to have reached their maximum extent in New Guinea shortly before 15,000 BP (Hope *et al.* 1976; Mercer 1983). Mercer (1983) tentatively proposed that the later advance of the tropical glaciers may be related in some way to the final major readvances of the northern hemisphere ice sheets prior to their rapid downwasting. It is possible, for example, that the relatively young moraine ages for the last glacial maximum may reflect the mass balances of many tropical glaciers that were largely influenced by changes in temperature rather than precipitation. Under such circumstances, the global warming that led to the melting of the last ice sheets may have begun worldwide between approximately 15,000 and 14,000 BP (Broecker and Denton 1990a,b).

Throughout North and South America, the Late Wisconsin mountain glaciation was associated with considerable ELA lowering, lower precipitation values and lower temperatures (Figure 4.7). For example, Porter *et al.* (1986) estimated that the Late Wisconsin ELA in the Cascade Ranges was lowered by 850–900 m and was associated with lower accumulation-season precipitation than present with mean annual temperatures 4°C lower than present. Similarly, Hamilton and Thorson (1983) estimated a Late Wisconsin ELA lowering for the Alaska Range of at least 200 m. By contrast, Clapperton (1983) observed that ELA lowering along the Andean mountain chain during the last glaciation varied from 400 m to a maximum of 1,000 m due principally to the effect of lowered temperatures. Clapperton (1987) also observed that in the Ecuadorian Andes during the last glaciation there may have been a mean annual temperature decrease of 7°C and an ELA lowering of between 860 m and 1,040 m. Similarly in Austria, and elsewhere throughout central and southern Europe, an ELA lowering during the last glacial maximum of between 1,000 and 1,200 m has been inferred (Figure 4.7) (Andersen 1981). Broecker and Denton (1990b) proposed that this broad similarity of mountain snowline depression across virtually all of the Earth's climatic zones was caused by high-mountain temperature lowering of nearly equal magnitude (between 4.2 and 6.5°C) in both polar hemispheres and the tropics. They argued that such a widespread pattern of temperature lowering could not be explained by Milankovitch mechanisms since these produced opposite insolation signals in both hemispheres.

It thus appears that there is fragmentary evidence for the occurrence of an early period of Late Quaternary glaciation that may have occurred during oxygen isotope stage 4. In the areas where a stage 4 glaciation seems to have taken place (e.g. in the Chilean Lake District and the French Pyrenees) ice expansion seems to have been more extensive than during the later isotope stage 2 glaciation. In many areas, the stage 2 ice advance appears to have culminated between 20,000 and 18,000 BP with a secondary glacial maximum having taken place in some areas between 16,000 and 15,000 BP. Such advances and retreats of mountain glaciers are likely to have been largely determined by variations in temperature rather than precipitation.

The volume of ice contributed by mountain glaciers and small ice caps during the last glacial maximum was equal to a global sea level change of approximately 5.5 m (Table 4.3). By contrast, it is estimated that world sea level may have fallen by as much as 117 m during the last glacial maximum if allowance is made for glacio-isostatic compensation. The various estimates of the areas, volumes and equivalent sea level thicknesses for the various mountain glaciers and small ice caps for both the present and the last glacial maximum have been summarised by Denton and Hughes (1981) and may be compared with estimates for the respective ice sheets (Table 4.3).

MODELS OF LATE QUATERNARY ICE SHEET EVOLUTION

Denton and Hughes (1981) proposed a remarkable scheme of Late Quaternary glaciation in which great emphasis was placed upon the role of marine-based ice shelves in affecting patterns of global glaciation and deglaciation. Their so-called 'outrageous' hypothesis has been strongly criticised, e.g. by Andrews (1982), who has succinctly described the main areas of controversy. The rationale of the Denton and Hughes argument is based upon the view that because the oxygen isotope records indicate a global fall in sea level of more than 75 m between 124,000 BP and 115,000 BP and because existing glaciological models cannot 'generate' the necessary ice volume in the time required, extensive ice shelves must have developed during this period in Hudson Bay, the Gulf of Bothnia, the Canadian Arctic archipelago and the Arctic Ocean. Thereafter, the ice shelves may have become grounded and enabled the growth of ice sheets. Andrews (1982) has strongly criticised this model and has shown that the field data (e.g. from Baffin Island) indicate a quite different model of Laurentide ice sheet growth in which ice growth took place initially on upland plateaux. The rejection or acceptance of the Denton and Hughes model is central to global models of glaciation and deglaciation during the Late Quaternary, especially with respect to the suggestion that a marine-based ice shelf covered the entire Arctic Ocean during periods of northern hemisphere glaciation.

This debate demonstrates quite clearly the way in which models of ice sheet growth and decay are developed alongside compatible or incompatible field data. Sometimes, the ice sheet models are not constrained by field evidence. Illustrative of this are models of the last ice sheet for northwestern Europe. In many areas, the limits of the last ice sheet occur offshore and, since most of these moraines are not dated, it is almost impossible to provide spatial constraints on the limits of the former ice sheet. Therefore, the ice sheet cannot be modelled easily – except on a hypothetical basis.

It is also clear that one ought not to expect different ice sheet margins to advance and retreat synchronously. In fact, one ought to expect the opposite since glaciers in high latitudes of the northern hemisphere appear to have been starved of precipitation during the last glacial maximum. Therefore, during a period of glaciation, ice advance along the southern margins of the Laurentide and Eurasian ice sheets was probably accompanied by ice retreat in high latitudes. This rather simplified picture is complicated by the occurrence of major ice surges that can take place regardless of climatic conditions. The message is clear: beware of simple explanations and assumptions of spatially synchronous behaviour – in most cases the answers are highly complex!

SUMMARY

It seems clear that the growth and decay of ice sheets in the northern hemisphere most probably took place on several occasions during the Late Quaternary. Thus the last interglacial was succeeded by a marked period of ice sheet growth in both North America and Scandinavia during isotope substage 5d that began at about 120,000 BP and culminated near 115,000 BP. An intriguing hypothesis is that this period of glaciation was indirectly related to the disintegration of the West Antarctic ice sheet and the occurrence of major ice surges near the end of substage 5e.

There is controversy about the manner in which ice built up during substage 5d. One view is that initial ice development took place on upland plateaux while another view is that the ice sheets developed from grounded ice shelves. The ice sheets that had formed during substage 5d were subject to widespread retreat and decay during substage 5c. Thereafter, renewed ice sheet build-up took place during substage 5b and, in North America, the ice may have been sufficiently extensive to have reached the St Lawrence Lowland.

Following a brief interval of interstadial warmth during substage 5a, extensive glaciation took place in the northern hemisphere during the transition between oxygen isotope stages 5 and 4 near 75,000 BP. The ice sheets that developed in the northern hemisphere at this time may have exceeded in size those that formed during isotope stage 2 and may locally have been nourished by moisture from the relatively 'warm' North Atlantic and North Pacific Oceans. Whatever the cause of ice sheet build-up at this time, the ice sheets lasted for only 15,000 years and had virtually disappeared by 60,000 BP at the end of isotope stage 4.

The periods of interstadial warmth that occurred during isotope stage 3 represented climatic fluctuations superimposed upon a pattern of global climate that had already switched to a glacial mode (Berger 1979). However, the occurrence of numerous pronounced Dansgaard–Oeschger oscillations in the Greenland ice core record also implies that there may have been several marked periods of climate cooling in the northern hemisphere during this isotope stage (see Chapter 3). The evidence for the first of the intervals of climatic warming, at 60,000 BP, although well represented in Europe, does not seem to have been such a prominent event in North America and as such its palaeoclimatic significance remains unclear. In Europe, it seems likely that this period of warmth lasted until approximately 55,000 BP when renewed growth of large ice sheets again took place in the northern hemisphere. For example, in western Norway, ice extended beyond the present coast for almost 25,000 years until near 33,000 BP.

There is some evidence from Europe and North America that a brief period of interstadial warmth occurred near 30,000 BP. Evidence from several areas (e.g. the USSR and eastern Canada) seems to imply that very severe cold conditions prevailed between approximately 42,000 BP and 35,000 BP before a well-defined period of interstadial warmth that occurred between approximately 33,000 BP and 25,000 BP. During this period, the northern hemisphere ice sheets contracted in size due to climatic warming. The last phase of ice sheet growth took place during isotope stage 2 and probably commenced soon after 25,000 BP. However, the culmination of the last glaciation appears to have occurred at different times in different parts of the world. For example, the glacial maximum of the Cordilleran ice sheet as well as many tropical glaciers may not have taken place until near 15,000 BP while many sections of the Laurentide ice sheet may have reached their maximum positions as early as 20,000 BP although some sectors continued to advance/surge southward long after this date.

Thus there is reason to believe that the growth and decay of large ice sheets in the northern hemisphere took place during the Late Quaternary on at least four and possibly five separate occasions. The Late Quaternary oxygen isotope stratigraphy from both ocean and ice cores illustrates clearly the rapidity, complexity and high magnitude of some of these changes. However, oxygen isotope stratigraphy, since it provides a measure of global ice volume changes over time, cannot provide information about where individual ice masses built up and decayed. At present, the greatest unknown factor is the history of the West and East Antarctic ice sheets throughout the Late Quaternary. It would be very interesting to discover if changes in volume of the ice sheets in Antarctica took place in phase or out of phase with the growth and decay of ice sheets in the northern hemisphere. There is good reason to believe that the Antarctic ice sheets attained their maximum extent during isotope stage 2 at 20,000 BP. However, the chronology of earlier Late Quaternary ice fluctuations in Antarctica remains very much a mystery.

5

THE MELTING OF THE LAST GREAT ICE SHEETS

INTRODUCTION

After the last glacial maximum, atmospheric warming gradually led to the eventual disappearance of the Eurasian and Laurentide ice sheets as well as many smaller ice caps, ice fields and valley glaciers. These changes led to a complex global response. For example, the rate of glacio-eustatic sea level rise was dictated by the patterns of ice melting while the enormous influx of fresh water into the world's oceans led to substantial changes in global ocean circulation. Furthermore, the melting of the large ice sheets of the northern hemisphere led to a complete reorganisation of global atmospheric circulation.

In the northern hemisphere, the melting histories of the Laurentide, Cordilleran and Eurasian ice sheets were non-synchronous. Thus, whereas the Cordilleran ice sheet had largely disappeared by 10,000 BP, the final melting of the Laurentide ice sheet was not completed until near 6,500 BP. In Scandinavia, ice sheet deglaciation was not completed until about 8,000 BP while, farther east, the remainder of the Eurasian ice sheet had disappeared slightly earlier, near 9,000 BP.

In all of these areas, the patterns of ice melting were highly complex. In many areas, large ice surges periodically reversed the overall pattern of ice retreat. On one occasion at least, renewed growth of ice sheets appears to have resulted from global climatic cooling. During the overall thinning and retreat of the ice sheets, immense ice-dammed lakes were formed along the ice margins. These lakes were some of the largest ever to have existed on Earth and played a vital role in the environmental changes that accompanied deglaciation. Some have even argued that the quantity of glacial meltwater released by them was sufficient to cool the waters of the North Atlantic and thereby induce long-lasting changes in global climate. In this chapter, most attention is given to the deglaciation history of the Laurentide ice sheet since it accounted for between 60 per cent and 70 per cent of the ice that melted during the decay of the last great ice sheets and also because its melting had important consequences for global climate changes. The following pages also discuss other large ice sheets; the deglaciation histories of the many smaller ice caps, ice fields and valley glacier complexes are also of great interest, but for reasons of space are not considered here in detail.

DEGLACIATION OF THE EURASIAN ICE SHEET

Deglaciation of the Soviet Union

The controversy

Two contrasting models of Eurasian ice sheet deglaciation have been described by Grosswald (1980) and Velichko *et al.* (1984). Whereas Grosswald considers that the extent of Eurasian ice cover was very large, Velichko *et al.* have argued for a considerably more restricted ice cover (Figure 4.1). According to Grosswald (1980) the melting of the Eurasian ice sheet was accompanied by remarkable diversion of the proglacial drainage along the southern margin of the ice sheet. Thus, the pattern of proglacial drainage southwards was initially replaced by a flow of water from east to west. In this way, the southward drainage of meltwater into the Caspian and Black Seas was gradually replaced by the westward drainage of meltwater into the lowlands of northern Poland and Germany (Figure 4.1). Grosswald (1980) has suggested that, after 14,000 BP, the principal westward-flowing route of proglacial discharge along the southern margin of the Eurasian ice sheet resulted in the development of extensive fluvial terraces and channels (*Urstromtaler*) in northern Germany and Poland and the eventual discharge of most meltwater into the North Sea and Norwegian Sea (Grosswald 1980: 19). As deglaciation continued, ice sheet disintegration in the Ob–Yenisei Lowlands resulted in the reversal in drainage of these rivers from their earlier southward courses to a pattern of northward flow into the developing Arctic Ocean.

The model of deglaciation proposed by Velichko *et al.* (1984) does not include an ice barrier in the Ob and Yenisei Lowlands, and hence in their model these rivers continued to flow into the Arctic Ocean throughout the Late Valdai/Sartan glaciation. The Velichko *et al.* (1984) model also envisages a limited ice cover over the Barents and Kara Sea shelves and mountain glaciation over Taimyr and Severnaya Zemlya (Figure 4.1). The Grosswald model of glaciation implies that most of the southern margin of the Eurasian ice sheet terminated in freshwater proglacial lakes. Grosswald (1980) depicts at least six moraine systems related to the limit of the Late Valdai/Sartan glaciation that he attributes to ice surging into adjacent water bodies (see Chapter 4).

Pattern of events

The timing of the initiation of deglaciation is not known with certainty although by 16,000 BP considerable ice melting and diversion of proglacial drainage had taken place. During deglaciation, several extensive moraines were deposited inside the outermost moraines (Figure 5.1). In European USSR, these include the Vepsovo, Krestzy, Luga, Neva and Palivere moraines that record a progressive ice margin retreat from south to north towards the White Sea. The ages of these moraines (Vepsovo the oldest at 15,000 BP and the Palivere moraine youngest at

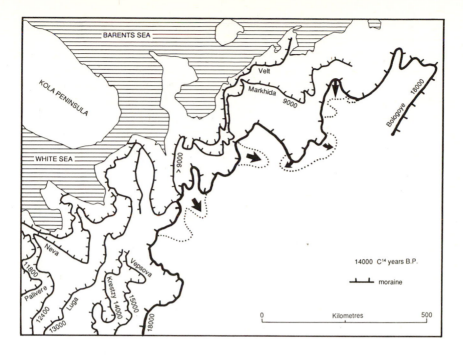

Figure 5.1 Late Weichselian/Valdai and Holocene ice-marginal moraines in northern European USSR showing former ice flow from north to south. Stippled outer lobes indicate possible areas of former ice surging (after Grosswald 1980).

11,800 BP) demonstrate that, by 11,800 BP, the remainder of the Eurasian ice sheet over European USSR was virtually severed from the Fennoscandian ice to the west (Figure 5.1).

During this period, regional deglaciation was also taking place in the region of the Barents Sea and Kara Sea ice domes on the Soviet Arctic continental shelf. The moraine sequence along the northern margin of the Late Valdai ice sheet is, however, not well dated although Grosswald (1980: 8) has proposed that widespread moraine deposition took place between 15,000 and 14,000 BP. A major ice advance is indicated by the Markhida and Velt moraines that extend for over 500 km across northwest USSR (Figure 5.1). These moraines appear to have been deposited at approximately 9,000 BP and are thought to represent widespread ice sheet surging from north to south during the final melting of the Eurasian ice sheet over the Barents Sea. Kind (1972), however, maintained that the Markhida moraines also reflected a climatically induced ice sheet readvance that interrupted the overall pattern of ice sheet decay. Kind argued that this climatic deterioration and ice advance (the Norilsk Stage) took place throughout central Siberia between 11,000 and 10,000 BP and was broadly equivalent in age to the Younger Dryas climatic deterioration generally agreed to have occurred

75

throughout NW Europe at this time (see p. 78). The age and location of the Markhida moraines indicate that complete deglaciation of the Eurasian ice sheet could not have been completed until after 9,000 BP. Evidence of a pronounced advance of Younger Dryas ice has also been proposed for the Spitzbergen archipelago (Boulton 1979). Boulton suggested that this advance (the Billefjorden Advance) may even have been associated with a more extensive ice cover than that which existed during the Late Weichselian maximum.

The large-scale glacio-isostatic depression (and later uplift) that was associated with the ice sheet (Grosswald estimates maximum dome thicknesses between 1.9 and 3.3 km and thus glacio-isostatic depression between circa 0.6 and 1.1 km) would have almost certainly permitted the penetration of marine and lake waters beneath parts of the stagnating Late Valdai/Sartan ice masses. Under such circumstances the scale of ice disintegration, ice surging and fluvioglacial discharge is likely to have been extensive. Arkhipov (1984) has speculated that the severing of the western USSR and Fennoscandian ice sheets during deglaciation may have led to the temporary establishment of a marine connection between the White Sea and the Baltic. However, detailed field mapping and stratigraphic studies undertaken by Hyvarinen (1973) and Eronen (1983) have shown that between the White Sea and the Gulf of Finland, ice decay was associated with the development of numerous large lakes and that the extent of marine sedimentation was considerably less than that proposed by Arkhipov (1984).

The deglaciation of Scandinavia

Early events

In general, it has proved difficult to establish an accurate deglaciation chronology for much of western Scandinavia. This is largely due to the discontinuous nature of the moraines and consequent difficulties of correlating between moraines in different fjord regions (Marthinussen 1974; Sollid et al. 1973; Andersen 1979) (Figure 4.3). An additional problem is uncertainty over the western extent of the ice sheet during the Late Weichselian glacial maximum (see Chapter 4).

The ages of the submerged end moraines on the continental shelf areas of western Norway are not known although the prevailing view is that the Egga I moraines may represent the limit of the last (Late Weichselian) ice sheet (Andersen 1979). Pertinent to this argument are radiocarbon dates of 18,400 BP from organic sediments from the island of Andøya (Figure 4.3) (Vorren 1978). According to Vorren, the Late Weichselian ice margin was located east of the island during the last glacial maximum. Since the Egga I moraines lie close to Andøya, it was inferred that these represent the limit of the last ice sheet in this region. More recently, Vorren et al. (1988) have suggested that the maximum advance of Late Weichselian ice in this region culminated between 19,000 BP and 18,500 BP and that the ice remained close to its maximum position until

16,000 BP. Thereafter, climatic amelioration took place between 16,000 BP and 13,700 BP when climatic cooling resumed.

Events after 13,000 BP

In Scandinavia, the chronology of Lateglacial environmental changes after 13,000 BP has been subdivided into an initial phase of interstadial warmth between 13,000 and 12,000 BP (the Bölling interstadial); a later brief phase of relatively cold stadial conditions between 12,000 and 11,800 BP (the Older Dryas Stadial); and another phase of interstadial warmth between 11,800 and 11,000 BP (the Alleröd interstadial) (Mangerud *et al.* 1979). This was followed by a severe climatic deterioration and ice advance during the Younger Dryas that ended at 10,000 BP (Figure 5.2).

These changes are well illustrated from the Bergen area of SW Norway where Mangerud *et al.* (1979) demonstrated that, following a short-lived Older Dryas ice sheet advance, rapid ice recession and thinning took place during the Alleröd and resulted in the inland retreat of the ice margin at least 50 km east of Bergen (Figure 5.2). Thereafter ice readvance commenced at 11,000 BP and continued until 10,100–10,000 BP when widespread retreat and stagnation took place.

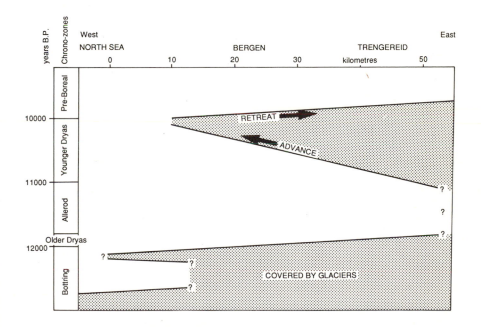

Figure 5.2 Reconstructed Lateglacial fluctuations of the Scandinavian ice sheet margin for the Bergen–Voss area (after Mangerud *et al.* 1979). Note the prominent readvance of the ice sheet during the cold climate of the Younger Dryas.

This advance of the ice sheet is recorded in Bergen by the Herdla moraines and is represented elsewhere in Scandinavia and in Iceland by well-defined end moraines (Grosswald 1984).

In southern Scandinavia, the initial ice-marginal retreat and thinning appears to have been interrupted by two separate ice advances into Denmark between 15,000 and 13,000 BP while ice-marginal recession reached the Swedish coast by 14,000 BP (Lundqvist 1986a). During this period, deglaciation of southwest Sweden took place while the southern Baltic between Sweden and Poland remained ice covered. The recession of the Late Weichselian ice sheet in southern Scandinavia appears to have been interrupted by ice readvances until 11,800 BP. Thereafter, accelerated ice retreat appears to have been a characteristic feature until 11,000 BP. Throughout this period, the areas of stagnating ice were surrounded in some places by freshwater lakes and in other places by the sea, the presence of the latter due to the complex interplay between glacio-isostatic land uplift and regional sea level rise.

Younger Dryas glaciation and the Baltic Ice Lake

Continued ice sheet decay in conjunction with glacio-isostatic uplift eventually resulted in the development of an extensive ice-dammed lake (the Baltic Ice Lake) that bordered the Fennoscandian ice sheet along much of its southern margin (Figure 5.3). The lake was initially created between southern Sweden and Poland between approximately 12,300 and 11,800 BP. However, continued ice sheet decay resulted in its extension by 11,000 BP into the area presently occupied by the Gulf of Finland. The Baltic Ice Lake was subject to a complex series of lake level fluctuations (Eronen 1983) (see Chapter 8). It also exerted an important influence on the pattern of deglaciation of southern Scandinavia since the changing position of the ice margin was controlled not only by climatic influences but also locally by calving into lake waters. A major Younger Dryas ice readvance took place into the lake soon after 11,000 BP after which time rapid and widespread deglaciation led to the final drainage of the lake at approximately 10,450 BP (see Chapter 8).

In western and northern Norway, the Younger Dryas moraines occur both upon land and also as submerged features in many fjord areas (Figure 4.3). In southern Norway, the limits of the Younger Dryas ice sheet are indicated by the well-known Ra moraines while in northern Norway the limit is represented by the Tromso–Lyngen moraines. In southern Sweden, the position of the former ice margin is represented by the Middle Swedish end moraines while in southern Finland they are continued as the Salpausselka moraines. The latter 'moraines' consist largely of deltas and minor drift ridges and are believed to have been deposited in close proximity to the Baltic Ice Lake. The Salpausselka moraines occur within three separate zones and indicate three former positions of the southern Fennoscandian ice sheet in Finland (Plate 3). One view is that the outermost moraine, Salpausselka I, represents the most advanced position of the ice during Younger Dryas time having been deposited between 11,200 and 10,950 BP (Donner and Eronen 1981).

According to Eronen (1983), ice recession and the formation of the Salpausselka II moraines (10,450–10,250 BP) was associated with the drainage of the Baltic Ice Lake (dated by varve chronology to 10,163 BP) while the inner moraines (Salpausselka III) were deposited just prior to the period of widespread ice retreat and stagnation (Eronen 1983). A contrasting view has been suggested by Bjorck and Digerfeldt (1986) who have argued on several grounds that the Baltic Ice Lake drained across the Middle Swedish end moraine zone on two separate occasions during the Lateglacial. The first drainage event took place near 11,250 BP and resulted in a lake level fall of between 10 and 15 m. Thereafter, ice advance during the Younger Dryas blocked the drainage outlet at Mt Billingen until 10,450 BP when a second (possibly catastrophic) drainage took place (Figure 5.3). The eventual drainage of the lake into the North Sea Basin at the close of the Younger Dryas may also have substantially affected the oceanography and seismicity of northwest Europe.

By 10,100 BP, ice sheet recession resulted in the establishment of a marine connection across southern Sweden between the Baltic and the North Sea forming the Yoldia Sea (Figure 5.3). Thus, by 10,000 BP, the southern margin of the Fennoscandian ice sheet was bordered by saline water from near Oslo in the west to Finland in the east. Thereafter, ice sheet disintegration continued and may not have been completed over Scandinavia until after 8,000–8,500 BP. Throughout this period, the interaction of glacio-isostatic uplift and regional eustatic sea level rise resulted in an alternating sequence of freshwater lake development and saline incursions in the Baltic.

The deglaciation of the British Isles

Introduction

Any discussion on the deglaciation of the British Isles should be prefaced by a statement that the extent of ice during the last glacial maximum is not known with any certainty. One view is that the last ice sheet was confluent with Fennoscandian ice over the North Sea and that the western margin of the ice sheet lay along the continental shelf (Boulton et al. 1977). A quite different view is that the last ice sheet in the British Isles was relatively restricted in its size and that its eastern margin was separated from the Fennoscandian ice sheet by dry land and areas of shallow water (Sissons 1981a) (Plate 2). Strong support for the latter view has recently been provided through detailed studies of the Late Quaternary stratigraphy of the North Sea (e.g. Sejrup et al. 1987). Accordingly, the patterns of ice sheet deglaciation that have been proposed reflect this difference of view. More recently, Eyles and McCabe (1989) have argued that the last British ice sheet, due to glacio-isostatic depression, was surrounded by extremely high sea levels (up to +150 m) that led to the widespread deposition of glaciomarine sedi-

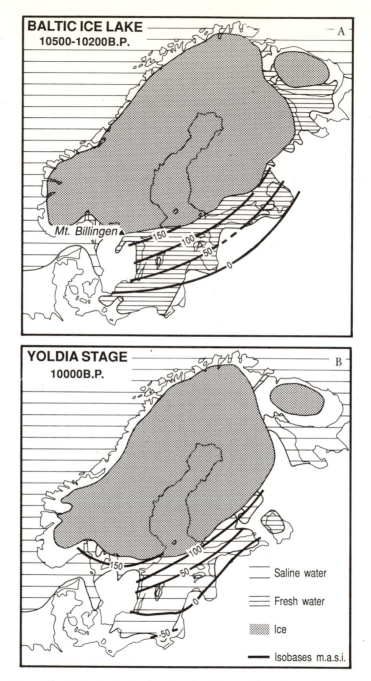

Figure 5.3 A. The Baltic Ice Lake showing shoreline uplift isobases (in metres) and the location of the lake drainage outlet near Mt Billingen. B. The extent of the Yoldia Sea at circa 10,000 BP showing the shoreline isobases (after Eronen 1983).

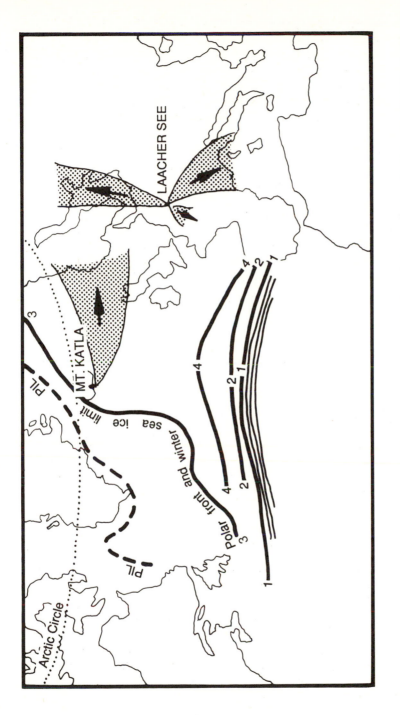

Figure 5.4 Position of the Polar Oceanic Front and the limit of winter sea ice during the period between circa 20,000 and 10,000 BP. 1. 20,000–16,000 years BP, 2. 16,000–13,000 BP, 3. 13,000–11,000 BP and 4. 11,000–10,000 BP. The thin lines depict the steep thermal gradient to the south of the Polar Oceanic Front. The approximate modern southern limit of sea ice is also shown (PIL). The ash plumes associated with the Mt Katla (Vedde ash) and Laacher See volcanic eruptions are also shown (see Chapter 10) (after Lowe and Walker 1984).

ments. The occurrence of high sea levels, in turn, led initially to the development of tidewater glaciers in the Irish Sea Basin and eventually to the collapse of the ice mass in this area. This is a minority view at present with most authors being of the opinion that regional deglaciation was associated with relative sea levels no higher than +40m around the retreating ice sheet margin.

It is generally agreed, however, that the melting of most ice was completed by 13,000 BP (Sissons and Walker 1974). Lowe and Walker (1984) noted that the greater part of this ice sheet stagnated while Britain remained surrounded by polar water and therefore concluded that deglaciation took place largely because of precipitation starvation, rather than an increase in temperature (Figure 5.4). Increased temperatures did, however, occur during the ensuing Lateglacial interstadial between 13,500 BP and 11,000 BP and prior to the cold climate of the Younger Dryas (Loch Lomond Stadial). The period of interstadial warmth was accompanied by a northward shift in the positions of the polar atmospheric and oceanic fronts in the North Atlantic region (Figure 5.4). Coope and Pennington (1977) have estimated that summer temperatures during the interstadial may have increased to values similar to present. However, the synoptic climatology of Britain during this period remained greatly influenced by the presence of permanent anticyclonic circulation over the Fennoscandian ice sheet and greater seasonality due to the occurrence of perihelion during summer rather than winter. Thus, in Britain, as elsewhere throughout the northern hemisphere, solar radiation was 7 per cent higher than present during summer and 7 per cent lower than present during winter (Rind *et al.* 1986).

The Lateglacial interstadial of the British Isles contrasts with that of southern Scandinavia where interstadial warmth is considered to have been interrupted by a brief period of cold conditions during the Older Dryas (Figure 5.2). In Britain, evidence for an Older Dryas climatic deterioration is only slight and possibly only local in effect (Lowe and Walker 1984). As a result, the time period comprising the Alleröd, Older Dryas and Bölling chronzones of southern Scandinavia is considered as broadly equivalent to one interval of relative warmth – the Lateglacial interstadial.

Younger Dryas glaciation

During the later part of the Lateglacial interstadial, a renewed southward migration of the polar oceanic front in the North Atlantic was accompanied by widespread ice accumulation in upland Britain during the Younger Dryas (termed the Loch Lomond Stadial in the British Isles). During this period, an ice cap up to 400 m thick developed in the western Highlands of Scotland while valley glaciers also developed in the English Lake District, North Wales and SE Ireland (Figure 5.5). Detailed mapping of former glaciers by Sissons (1979a) has enabled the reconstruction of equilibrium line altitudes for this period of glacier expansion. These have been used to estimate former mean July temperatures of between 6 and 8°C at sea level in northern Britain and typical mean January temperatures of

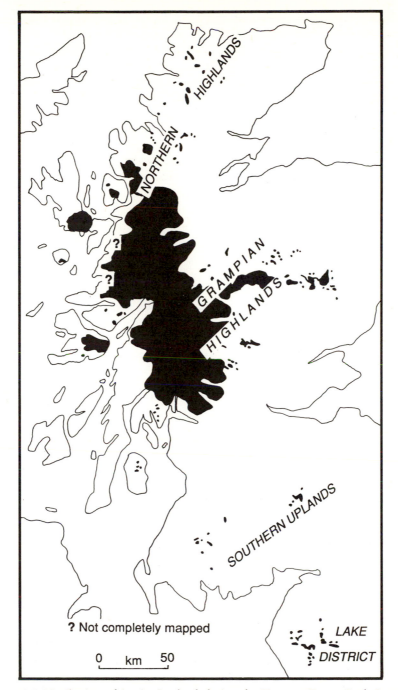

Figure 5.5 Distribution of ice in Scotland during the Younger Dryas (Loch Lomond Advance) (after Sissons 1974a).

between −17 and −20°C; low enough for the development of permafrost.

The reconstructed temperature values indicate the enhanced seasonality of the climate caused by the reversal of perihelion and aphelion (compared to the present). However, the values also indicate a climatic deterioration of great severity where the precipitation regime was dictated by the changing position of the polar atmospheric and oceanic fronts in the North Atlantic. The highest equilibrium line altitude (ELA) values (up to 1,000 m) during the Younger Dryas appear to have occurred in the Cairngorm Mountains, eastern Scotland, and may have been as low as 300 m in western Scotland. During this period, a more continental climate may have prevailed over Britain indirectly related to the occurrence of dry land over the present central and southern North Sea areas. As a result, cold easterly winds that originated from the southern areas of the Fennoscandian ice sheet anticyclone are likely to have resulted in limited snowfall yet very low air temperatures over eastern Britain. By contrast most snow precipitation probably resulted from SE and SW winds associated with cyclones that tracked to the east and northeast along the polar atmospheric front in the North Atlantic and across the British Isles.

The timing of Younger Dryas glacier expansion is not known with certainty and it is possible that considerable ice accumulation may have taken place prior to 11,000 BP. Lowe and Walker (1984) have argued that the Younger Dryas glacial maximum may have occurred around 10,800 BP while widespread glacier retreat, in contrast with Scandinavia, may have taken place as early as 10,500 BP.

DEGLACIATION OF THE LAURENTIDE AND CORDILLERAN ICE SHEETS

The Laurentide ice sheet

Introduction

The concept of a very extensive Late Wisconsin Laurentide ice sheet (e.g. Flint 1971) has in recent years been replaced by the suggestion that, with the exception of the American Midwest, the Late Wisconsin ice sheet was much smaller than previously considered (Dyke and Prest 1987) (see Chapter 4). There is general agreement on the extent of ice expansion into the American Plains and also the Atlantic Provinces (Denton and Hughes 1981; Clayton and Moran 1982; Mickelson *et al.* 1983; Grant 1987, 1989; Fulton 1989) although there are different views on the extent of Late Wisconsin ice in the Canadian Arctic (Hughes *et al.* 1981; Andrews *et al.* 1986).

Controversy also exists on the surface geometry of the ice sheet, with a continued debate on the positions of the major ice divides, domes, major saddles, ice streams and ice shelves (Dyke and Prest 1987). Flint (1943) and more recently Mayewski *et al.* (1981) have argued that the Laurentide ice sheet was dominated by a single dome that flowed radially outward over Hudson Bay. However, many authors (e.g. Shilts *et al.* 1979; Shilts 1980; Andrews 1982; Dyke and Prest 1987) argue that the distribution of glacial striae and erratics indicates that the patterns of ice flow were probably much more complex reflecting outward ice dispersal from independent ice centres over Labrador, Keewatin and the Foxe Basin (see Chapter 4). Finally, there is disagreement on the timing of ice marginal fluctuations. It is generally believed, for example, that the principal ice margins fluctuated out of phase with each other and that a glacial advance in one area might have taken place at the same time as ice retreat in another.

The significance of ice surging

The enormous glacio-isostatic depression associated with the Laurentide ice sheet resulted in the development of proglacial lakes along many areas of the southern ice sheet margin. The presence of these lakes may have been of great importance in determining the conditions (e.g. yield stresses of subglacial sediments) beneath the ice sheet and may have thus been of critical importance in the surging of ice lobes through the reduction in basal shear strength beneath the ice (Clayton *et al.* 1985). The patterns of ice flow are further complicated by the likelihood that many areas of the Laurentide ice sheet developed upon pre-existing permafrost that had been produced during the Middle Wisconsin non-glacial interval (Fisher *et al.* 1985).

It has been suggested that permafrost degradation by geothermal heating may have permitted ice advance over a deformable substrate and hence provided an environment suitable for major surging of ice lobes (Fisher *et al.* 1985). Consequently, it has proved extremely difficult to distinguish former ice advances caused by climatic deterioration from those attributable to surging. The occurrence of former large-scale ice surging over deformable beds also alters ice sheet surface profiles as well as the ice mass balance which, in turn, theoretically alters the position of ice sheet divides and the overall flow behaviour of the ice sheet. Equally the transfer of ice from high to low elevations is likely to have locally increased calving into lakes and to have increased net ice ablation.

The deglaciation history of the Laurentide ice sheet appears to have been characterised by a large number of major ice surges (see below). In many cases, lobes of ice surged into ice-marginal lakes. Until recently many of the ice surge events were not recognised as such and were instead considered indicative of ice advances due to climatic cooling. At present, there are no ice readvances associated with the decay of the Laurentide ice sheet that are unequivocally related to periods of climatic cooling. The only possible exception is a period of ice advance in the Atlantic Provinces that may have taken place during the Younger Dryas.

Early fluctuations of the ice sheet margin

The southern margin of the Laurentide ice sheet was dominated by a series of ice lobes that are believed to have advanced and retreated non-synchronously (Mick-

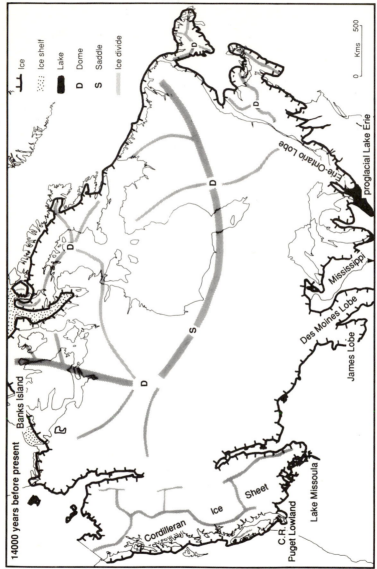

A

14000 years before present

Banks Island

Cordilleran

Ice

Sheet

C.R.
Puget Lowland

Lake Missoula

James Lobe

Des Moines Lobe

Mississippi

Erie-Ontario Lobe

proglacial Lake Erie

D

S

D

D

D

D

D

Ice

Ice shelf

Lake

D Dome

S Saddle

Ice divide

0   Kms   500

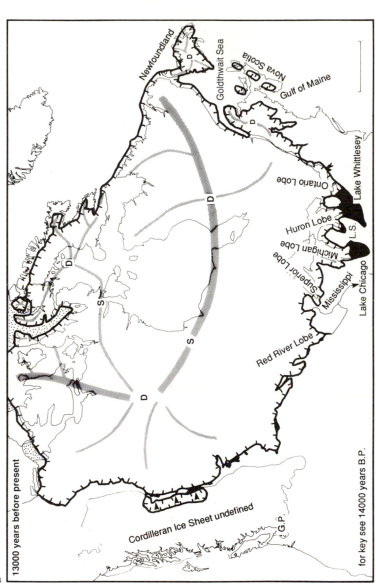

B

13000 years before present

Newfoundland

Goldthwait Sea

Nova Scotia

Gulf of Maine

Lake Whittlesey

Ontario Lobe

L.S.

Huron Lobe

Michigan Lobe

Superior Lobe

Mississippi

Lake Chicago

Red River Lobe

Cordilleran Ice Sheet undefined

G.P.

for key see 14000 years B.P.

Figure 5.6 Reconstruction of the Laurentide ice sheet for A. 14,000 BP and B. 13,000 BP showing positions of major ice divides. C.R. = Columbia River, G.P. = Glacier Peak, L.S. = Lake Saginaw (after Dyke and Prest 1987; reprinted with permission of the Minister of Supply and Services Canada).

elson *et al.* 1983). In Minnesota, Iowa and South Dakota the major ice accumulations of the Des Moines and James Lobes (Figure 4.4) advanced southwards
into the headwaters of the Mississippi drainage basin. Farther east, major lobes
advanced over the Green Bay and Lake Michigan Basins while in the southern
Great Lakes region, several connected lobes advanced into central Indiana and
Ohio. In Ohio, Pennsylvania and New York States, the advancing ice lobes were
confined by the Appalachian Plateau. Farther east in the New England States and
the Gulf of Maine, the predominant ice flow was from north to south and
according to some authors extended south of Cape Cod and as far as Long
Island. According to Prest and Grant (1969) the ice extended onto the Scotian
Shelf where it formed a marine-based ice shelf (Figure 4.4).

During the advance of the Late Wisconsin ice sheet into Manitoba and
Ontario, the northward-draining rivers of central Canada and north-central
United States were impounded to produce a large lake. The lake, which initially
overflowed to the SE across Ontario towards the eastern Great Lakes, was
progressively displaced southwards due to the advance of Late Wisconsin ice
(Teller and Clayton 1983). The lake overflow carried considerable volumes of
meltwater into the Mississippi drainage basin but by 20,000 BP, the ice of the
Red River Lobe had again advanced and finally extinguished the lake. Thereafter,
several ice fluctuations of the Red River Lobe resulted in the alternate formation
and extinction of lakes with the final ice advance of the Des Moine Lobe into
Iowa culminating at 14,000 BP (Figures 5.6A and B). These lakes were the prede-
cessors of glacial Lake Agassiz (Teller and Clayton 1983) which again began to
form during ice sheet deglaciation after 12,300 BP (see p. 93).

Non-synchronous ice-marginal fluctuations

The timing of deglaciation along the southern margin of the Laurentide ice sheet
is complicated by the occurrence of regional ice advances, often major ice surges,
that interrupted the overall pattern of ice thinning and retreat. The most striking
characteristic, however, is that the western and southwestern margins of the
Laurentide ice sheet were subject to early retreat whereas the eastern area of the
ice sheet may have remained dynamically active until as late as 14,000 BP (Figure
5.6A). This contrast is attributed by Dyke and Prest (1987) to increased aridity
over the western sectors of the Laurentide ice sheet consequent upon the build-up
of the Cordilleran ice sheet and upon the establishment of permanent anticyclonic circulation above the ice sheet surface. Over the eastern areas of the ice
sheet, the convergence of jet streams over the Davis Strait and Labrador Sea may
have provided conditions favourable for cyclone development and snow precipitation.

Thus, progressive ice thinning and retreat took place in South Dakota, North
Dakota and Montana after 17,000 BP. By contrast, ice advances over parts of
southern Minnesota, central Iowa and southern South Dakota took place as late
as 14,000 BP and mark the maximum extent of Late Wisconsin ice advance in

this area (Mickelson *et al.* 1983) (Figure 5.6A). Similarly, the James Lobe in South Dakota was associated with considerable ice advance (circa 800 km), possibly resulting from a surge (Clayton *et al.* 1985), that culminated at 14,000 BP (Figure 5.6A).

In the southern Great Lakes region, in Illinois, Indiana and Ohio, the overall pattern of ice retreat was interrupted by numerous readvances of individual ice lobes. The enormous scale of some of the ice-marginal fluctuations is shown, for example, by the readvance of the Lake Huron ice lobe at 15,500 BP which may have advanced more than 600 km (Evenson and Dreimanis 1976). By contrast, major ice retreat took place in Ohio prior to 15,000 BP when the Erie–Ontario Lobe retreated from close to the Late Wisconsin ice limit to the Ontario Basin resulting also in the development of proglacial Lake Erie (Mörner and Dreimanis 1973).

Ice sheet fluctuations in the Great Lakes region

The period between 15,000 BP and 13,000 BP was primarily characterised by widespread ice thinning and retreat between eastern Wisconsin (the Green Bay Lobe) and the Lake Michigan Basin (the Lake Michigan Lobe) as well as throughout large areas of Michigan (the Saginaw Lobe) (Figure 5.6). Exceptions to this pattern are the James and Des Moines Lobes, which advanced up to 800 km reaching their Late Wisconsinan maximum position at 14,000 BP (Clayton and Moran 1982; Teller 1989) (Figure 5.6A). The decay of these ice masses preceded a major readvance of ice that appears to have culminated at 12,900 BP throughout the northern Great Lakes region. This ice advance (the Early Port Huron Advance) resulted in the development of a number of distinct lobes, the most important of which occurred between Lakes Erie and Ontario as well as two additional lobes along the southern margins of Lake Huron, while a major lobe still occupied northern Lake Michigan (Figure 5.6B).

The ice advances also resulted in the damming of large proglacial lakes: Lakes Whittlesey and Saginaw in the Huron–Erie Basins and Lake Chicago south of the Michigan ice lobe (Figure 5.6B). Lakes Whittlesey and Saginaw temporarily over-spilled into Lake Chicago which, in turn, overflowed into the Mississippi drainage basin. Throughout the region, the influence of proglacial water bodies on ice margin stability was considerable and undoubtedly played a major role in dictating the ice dynamics of individual lobes by reducing the subglacial yield stresses. It is not known whether the Early Port Huron Advance represents a regional climatic deterioration or whether the advance is attributable to ice surging although the prevailing view is for the latter (Dyke and Prest 1987) (Figure 5.6B). This is because it is extremely difficult to distinguish between glacigenic sediments attributable to (climatically induced) ice advances from those deposited as a result of ice surging.

The pattern of deglaciation in the Great Lakes region after the Port Huron Advance involved not only the formation of proglacial ice-dammed lakes, but

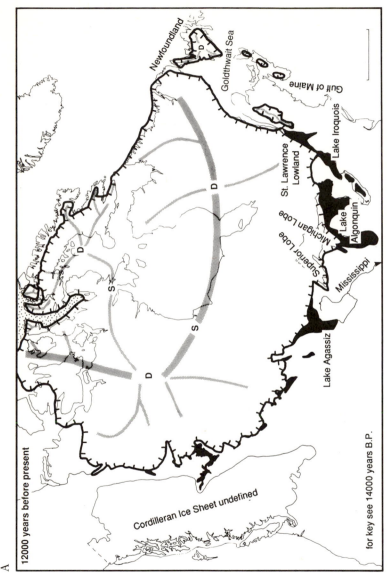

A

12000 years before present

Newfoundland

Goldthwait Sea

Gulf of Maine

Lake Iroquois

St. Lawrence
Lowland

Michigan Lobe

Lake
Algonquin

Superior Lobe

Mississippi

Lake Agassiz

Cordilleran Ice Sheet undefined

for key see 14000 years B.P.

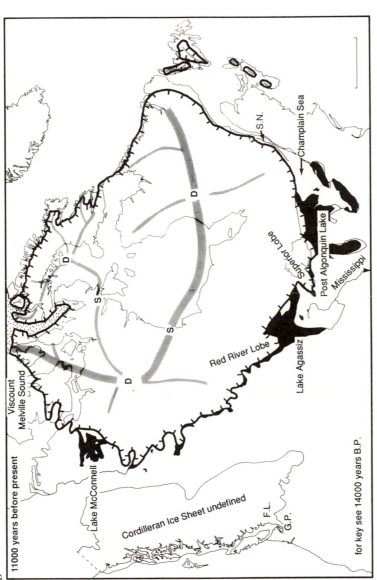

B

11000 years before present

Viscount
Melville Sound

Lake McConnell

Cordilleran Ice Sheet undefined

F.L.
G.P.
G.P.

for key see 14000 years B.P.

Red River Lobe

Lake Agassiz

Post Algonquin Lake

Mississippi

Superior Lobe

Champlain Sea

S.N.

D

S

S

D

D

Figure 5.7 Reconstruction of the Laurentide ice sheet for A. 12,000 BP and B. 11,000 BP showing positions of major ice divides. F.L. = Fraser Lowland, G.P. = Glacier Peak, S.N. = Saint Narcisse moraine (after Dyke and Prest 1987; reprinted with permission of the Minister of Supply and Services Canada).

also the displacement (sometimes through catastrophic jökulhlaups and some-times simply by the overflow from one lake to another) of large volumes of water. In general the ice sheet continued to thin and retreat and, by 12,000 BP, retreat of the Michigan and Huron Lobes resulted in the coalescence and expansion of Lakes Whittesley, Chicago and Saginaw into a new Lake Algonquin (Figure 5.7A). Farther east, stagnation and retreat of the Ontario ice lobe and of ice in the Quebec valley resulted in the formation of glacial Lake Iroquois (also supplied by Algonquin discharge) and the separation of the lake from the Cham-plain Sea to the northeast by a narrow tongue of ice (Figure 5.7A). By 11,000 BP, however, progressive deglaciation had resulted in the formation of the larger Post Algonquin Lake and its overflow into the Champlain Sea (Dyke and Prest 1987) (Figure 5.7B).

Local ice advances did still occur throughout this period. The most notable of these are advances over western and central New York State at 12,600–12,400 BP (Fullerton 1980; Mickelson et al. 1983) in the Lake Erie Basin and a major advance of the Lake Michigan and Green Bay Lobes as well as a considerable advance of ice west of Lake Superior (the Nickerson phase) (Figure 5.7A). In general, accelerated ice retreat and thinning took place along most of the Lauren-tide ice sheet margin with the exception of the eastern and northeastern areas between Ellesmere Island and Newfoundland (Dyke and Prest 1987). Retreat of the Nickerson phase ice margin may have commenced at 11,500 BP and, by 11,000 BP, the ice margin had retreated far enough north to permit the outflow of lake waters from Lake Agassiz into Lake Superior from its previous routing through the Mississippi River to the Gulf of Mexico (Figure 5.7B).

Water entered Lake Superior via the Lake Nipigon region as a series of catas-trophic jökulhlaups due to the repeated breaching of ice barriers that led to the widespread deposition of flood sediments (Teller and Clayton 1983; Teller and Mahnic 1988). According to Broecker et al. (1989), Broecker and Denton (1990a,b) and Teller (1990), these events had a profound influence on the climate of the northern hemisphere. They argue that at this time the flow of glacial meltwater from the Laurentide ice sheet along the Mississippi into the Gulf of Mexico was effectively stopped. Instead, meltwaters began to drain eastwards from Lake Superior along the St Lawrence and into the western North Atlantic. Enormous volumes of fresh water drained into the North Atlantic and, in doing so, increased the salinity stratification of the waters and thus inhibited the formation of deep water. Clayton (1983) and Teller and Thorleifsen (1983) argued that this process continued uninterrupted until 9,900–9,500 BP, when ice surging (the Marquette Advance – the last major ice surge in the Great Lakes region) along Lake Superior during the Emerson phase blocked the movement of water from Lake Agassiz into Lake Superior. Broecker et al. (1989) and Broecker and Denton (1990a,b) proposed that this massive influx of fresh water and the inhibition of deep water formation may have played an important role in causing the cold climate of the Younger Dryas (see below). If this is true, it provides a good illustration of the way in which glacier behaviour may have influenced global climate even though global climate change may have been primarily responsible for progressive ice sheet decay.

The development of glacial Lake Agassiz

The progressive retreat and thinning of the Red River Lobe eventually resulted, shortly after 12,000 BP, in the development of glacial Lake Agassiz (Figures 5.7 and 5.8). The overflow waters of Lake Agassiz drained first into the Minnesota River valley towards the Minneapolis region, where they joined the Mississippi. The sudden influx of glacial meltwater into the Gulf of Mexico during this period is clearly indicated in the Lateglacial oxygen isotope stratigraphy of the Gulf sea floor sediments where a 'spike' of increased ^{18}O depletion is evident for the time period between 12,500 and 12,000 BP (Broecker *et al.* 1989). Broecker *et al.* (1989) also noted, however, that a very pronounced decrease in ^{18}O depletion is evident in the Gulf of Mexico oxygen isotope record between 11,000 and 10,000 BP (Figure 5.8A inset). They concluded that this change was indicative of a reduced flux of glacial meltwater down the Mississippi River and observed that this time interval corresponds with the period when meltwater discharge was rerouted eastward through the Great Lakes and St Lawrence into the North Atlantic.

During its 4,000 year history, glacial Lake Agassiz played a fundamental role in the deglaciation history of the Laurentide ice sheet. Lake Agassiz reached its maximum extent between 9,900 and 9,500 BP when it covered an area of 350,000 km^2 and occupied large areas of Ontario and Manitoba. During its history, ice sheet thinning and retreat was interrupted on several occasions by major ice advances (Figures 5.8A and B). As previously noted, the lake over-flowed into the Minnesota River valley. However, by 10,700 BP retreat of the Lake Superior Lobe permitted the overflow of Lake Agassiz waters into northern Lake Superior (Clayton 1983). Subsequent advance of the Lake Superior Lobe at 9,900 BP closed the eastern outlet for Lake Agassiz and resulted in a lake level rise (the Emerson phase) and renewed lake overflow into the Minnesota valley (Figure 8.4). Thereafter a series of catastrophic jökulhlaup floods repeatedly breached the adjacent ice barriers and caused periodic and catastrophic flooding through the Nipigon Basin into Lake Superior and the St Lawrence. These floods may represent some of the largest floods ever to have occurred on Earth during the Quaternary (see Chapter 8).

The progressive increase in surface area of Lake Agassiz to a maximum between 9,900 BP and 9,500 BP may have provided an important source of mois-ture, and hence precipitation, for the mid latitude cyclones that tracked from west to east along the southern margins of the ice sheet. Of importance also is the development between 11,000 and 10,000 BP along the western Laurentide ice border of glacial Lake McConnell which was of comparable size to Agassiz (Figure 5.8A).

A

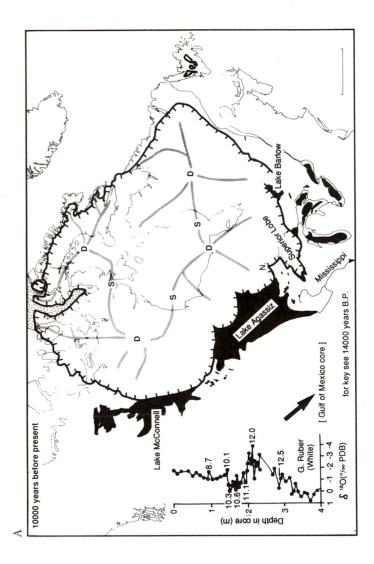

10000 years before present

Lake McConnell

Lake Agassiz

Lake Barlow

Superior Lobe

Mississippi

[Gulf of Mexico core]

for key see 14000 years B.P.

G. Ruber
(White)

8.7

10.1

10.3
10.6
11.1

12.0

12.5

Depth in core (m)

δ 18O (‰ PDB)

4   3   2   1   0   -1   -2   -3   -4

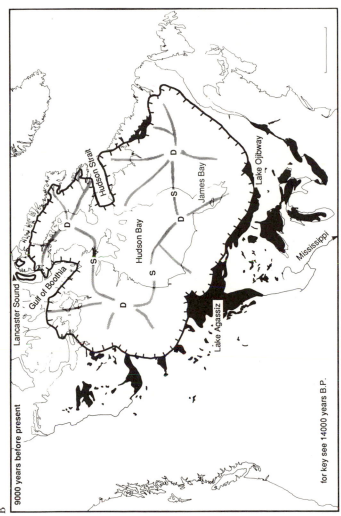

Figure 5.8 Reconstruction of the Laurentide ice sheet for A. 10,000 BP and B. 9,000 BP showing positions of major ice divides. The inset shows the ¹⁸O planktonic foraminifera record from Gulf of Mexico core EN32-PC4 showing a 'spike' due to Mississippi meltwater discharge. The high discharge ended just before 11,000 BP when meltwater was rerouted eastwards via the St Lawrence to the western North Atlantic Ocean. Resumed Mississippi drainage took place soon after 10,000 BP. The numbers are ¹⁴C ages in thousands of years BP N = Nipigon (after Dyke and Prest 1987 and Broecker *et al.* 1989; reprinted with permission of the Minister of Supply and Services Canada).

Ice sheet deglaciation in the Atlantic Provinces

According to Mayewski *et al.* (1981), the limit of Late Wisconsin ice in the Atlantic Provinces is considered to have extended south of Cape Cod and onto Long Island. However, an alternative view has been proposed by Grant (1977, 1989) who argued that the ice sheet was much smaller in dimensions and may not have extended beyond the present coasts of Maine and New Hampshire and that independent ice caps may have existed over Newfoundland and Nova Scotia.

The scheme of deglaciation proposed by Grant (1977, 1989) is one in which most ice masses were land based. Grant suggests that the maximum limit of Late Wisconsin ice in Newfoundland was reached about 12,700 BP and that subsequent ice decay was interrupted by two later glacial readvances. Implicit in the Grant model of Late Wisconsin glaciation and deglaciation of the Atlantic Provinces is that the Gulf of Maine remained essentially free of ice during this period and that the St Lawrence Lowland was free of ice as far west as the Saint Atonin interlobate moraine (northeast of Quebec). Certainly, during the preceding period between 18,000 and 14,000 BP the margin of the Laurentide ice sheet in the Atlantic Provinces and the Gulf coastal areas of Quebec remained relatively stable, contrasting strikingly with the retreating ice margin farther west. Dyke and Prest (1987) attributed this characteristic to a greater precipitation supply to the Atlantic sector of the ice sheet during this period. The latter authors also noted that there is no evidence of any major ice retreat between 18,000 and 14,000 BP for most of the Laurentide ice sheet margin between Nova Scotia and Banks Island in the Canadian Arctic (Figure 4.5).

A different account is given by Mayewski *et al.* (1981) who have argued that the deglaciation of the Atlantic Provinces from an initially extensive ice cover was dominated by the separation from the main ice sheet of an ice cap located between the Gulf of Maine and the St Lawrence Lowland. They considered that extensive ice calving and a relative marine transgression in the St Lawrence Lowland eventually resulted in the development of the Goldthwait and Champlain Seas. Richard (1978) has argued that this relative marine transgression may have reached Ottawa by 12,700 BP while others consider that the relative marine transgression took place slightly later (Dyke and Prest 1987). Farther east, rapid ice sheet calving and retreat may have taken place in the Gulf of Maine while on the adjacent land masses of Connecticut, ice sheet retreat took place more slowly.

In general terms, a large proportion of the deglaciation of the Atlantic Provinces took place at a relatively early stage (perhaps by 13,000 BP) in the overall melting of the Laurentide ice sheet. Certainly, most areas were ice free by 11,500 BP although periodic ice advances took place from the north into the Champlain Sea (LaSalle and Elson 1975). LaSalle and Elson have suggested that the St Narcisse moraine that trends for 300km along the north flank of the St Lawrence river valley was probably deposited in this manner at 11,000 BP (Figure 5.7B). Between 12,000 and 11,000 BP, however, the overall calving retreat of the ice sheet between Labrador and Lake Superior was much slower than in western areas where the ice margin was mostly land based (Dyke and Prest 1987). The most conspicuous ice advances during this period appear to have been surges. For example, several surges covering hundreds of kilometres took place into the Lake Agassiz Basin (Clayton *et al.* 1985; Klassen 1983), while in the Canadian Arctic a 60,000 km² surge took place into Viscount Melville Sound (Hodgson and Vincent 1984). A possible exception to this trend is an ice advance that culminated in Quebec and also in Newfoundland at 10,000 BP and which has been attributed to Younger Dryas cooling (Dubois and Dionne 1985; Grant and King 1984; Mott *et al.* 1986; Dyke and Prest 1987) (Figure 5.8A).

Deglaciation during the Late Pleistocene–Holocene transition

Introduction

The patterns of northern hemisphere atmospheric and oceanic circulation that accompanied the deglaciation of the Laurentide ice sheet during and following the Late Pleistocene–Holocene transition played a critical, yet still largely unknown, role in the progressive disintegration of the ice sheet. Of great importance to the decay of the Laurentide ice sheet was the fate of the two jet streams that had previously flowed to the north and south of the ice sheet and most probably merged in the lee of the ice sheet over the eastern North Atlantic (Broccoli and Manabe 1987). The inferred switch to the present mode of jet stream circulation must have occurred when both jet streams merged to descend over the American Great Plains between the Rockies and the stagnant Laurentide ice sheet. However, the timing of this change is not known.

Significant changes also occurred in the North Atlantic Ocean, where at 10,000 BP the polar oceanic front migrated northwards to close to its present position (Figure 5.4). Consequently, the disappearance of sea ice from the North Atlantic, the re-establishment of the Gulf Stream and the deglaciation of large areas of the Eurasian ice sheet greatly altered the synoptic climatology of the northern hemisphere. Of equal importance were the effects that the melting ice sheet had on the patterns of northern hemisphere climate during this period. The type of feedback effect, where glacier activity influenced climate, was principally related to the supply of large volumes of glacial meltwater to the North Atlantic and the formation of North Atlantic deep water (Broecker *et al.* 1989) (see Chapter 8). These mechanisms of altered oceanic and atmospheric circulation coupled with Holocene global warming underpin the patterns of Holocene Laurentide ice sheet deglaciation described below.

Progressive deglaciation and ice surges

According to Dyke and Prest (1987) significant retreat of Keewatin and Hudson ice had taken place by 9,000 BP (Figure 5.8B). However, the greatest retreat was of the Superior Lobe which retreated 600 km (following the preceding Marquette

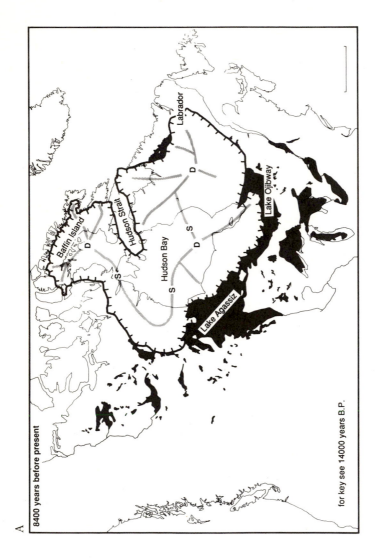

A

8400 years before present

Greenland

Baffin Island

Hudson Strait

D

S

Hudson Bay

D

S

S

D

Labrador

Lake Ojibway

Lake Agassiz

for key see 14000 years B.P.

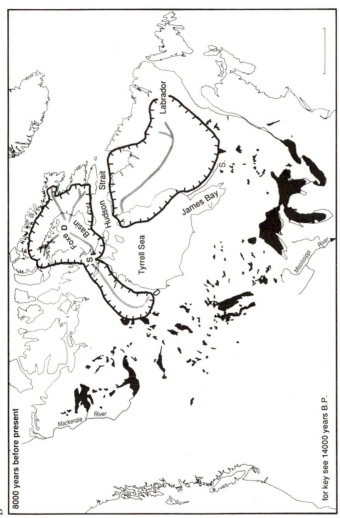

B

8000 years before present

Mackenzie River

Foxe D Basin

Hudson Strait

S

Labrador

Tyrrell Sea

James Bay

Mississippi River

for key see 14000 years B.P.

Figure 5.9 Reconstruction of the Laurentide ice sheet for A. 8,400 BP and B. 8,000 BP showing positions of major ice divides. The distribution of ice for 8,000 BP may also include a floating ice island (not shown) over Hudson Bay that was produced during the final disintegration of the Laurentide ice mass and the drainage of glacial Lake Agassiz/Ojibway through the Hudson Strait. S = Sakami moraine (after Dyke and Prest 1987; reprinted with permission of the Minister of Supply and Services Canada).

surge) and of the Gulf of Boothia and Lancaster Sound ice shelf in the Canadian Arctic which appears to have disintegrated and caused an ice marginal retreat of 1,000 km (Figure 5.8B). In the latter case, and elsewhere throughout the Canadian Arctic, rapid ice recession and disintegration was accelerated by a marine transgression along the ice margin and by a reduction in the ice supply (Dyke and Prest 1987). In the south, Lake Agassiz had merged with adjacent lakes to form a 1,800 km long ice-marginal lake that overflowed into the St Lawrence Lowland via the Ottawa valley (Figure 5.8B).

The pattern of progressive ice retreat and thinning of the ice sheet appears to have been interrupted at 8,400 BP by a series of major ice advances (the Cochrane Advances) throughout the Canadian Arctic, the Hudson Bay and James Bay Lowlands (Dyke and Prest 1987) (Figure 5.9A). Some of the moraines may represent an increase in snow precipitation (rather than a temperature decrease) caused by the frequent development of cyclones. However, considerable surging also occurred into adjacent freshwater ice-marginal lakes. For example, ice advanced to the southeast across the James Bay Lowlands into the adjacent Lake Ojibway (Vincent and Hardy 1979) (Figure 5.9A). The southern and SW ice sheet margin was thus flanked by Lakes Agassiz and Ojibway that had been connected, due to ice retreat, to create a 3,100 km lake margin (Dyke and Prest 1987).

Ice sheet disintegration and the catastrophic drainage of glacial Lake Agassiz–Ojibway

In general terms, the Cochrane ice advances were succeeded by the disintegration of ice in the Hudson Bay and James Bay Lowlands and the incursion into the Hudson Basin of the Holocene (Tyrrell) sea through the Hudson Strait (Figure 5.9A). Specifically, however, the disintegration of Hudson Bay and James Bay ice also resulted in probably the most catastrophic and dramatic sequence of events ever to have affected the northern hemisphere during the Late Quaternary. At 8,000 BP, ice sheet stagnation was accompanied by the catastrophic drainage of glacial Lake Ojibway and glacial Lake Agassiz into the Hudson Basin (Figure 5.9B). The breaching of the ice barrier that separated Lake Ojibway from the newly created Tyrrell Sea in the James Bay Lowland was accompanied by a fall in lake level as the combined Agassiz/Ojibway lake drained through the ice (Dyke and Prest 1987). Crude estimates indicate that between 70,000 and 150,000 km³ of water was drained from Agassiz and Ojibway during this period (see Chapter 8).

This event marked the final change in the route of large-scale meltwater drainage from the Laurentide ice sheet. The meltwater continued to drain into the North Atlantic but this time most of the drainage was through the Hudson Strait rather than the St Lawrence as it had previously been prior to 8,000 BP. Unlike the meltwater drainage that took place between 11,000 and 10,000 BP, the influence of the 8,000 BP drainage on global climate is not known. It remains to be discovered, for example, why the very large influx of freshwater into the North Atlantic at 8,000 BP did not influence NADW production and cause a renewed period of cold climate analogous to the Younger Dryas.

Crude estimates of the volume of water that drained from Lakes Ojibway and Agassiz at 8,000 BP indicate that catastrophic drainage of the water would have resulted in a virtually instantaneous raising of world sea level by between circa 0.2 and 0.4 m within 2 days.[1] The tectonic implications of such a rapid addition of mass over the world's oceans and the removal of such a water load from the area previously occupied by Lakes Agassiz and Ojibway have not been considered previously. It seems inevitable, however, that significant seismicity and earthquake activity may have occurred in the central Canadian Shield region.

Throughout this period of ice disintegration, comparatively large volumes of ice remained over Labrador and Baffin Island. Collapse of the southern ice margin during the lowering of Lake Ojibway resulted in the development of a 600km long moraine (the Sakami moraine) in Quebec east of James Bay (Hillaire-Marcel et al. 1981). After 8,000 BP much of the ice retreat was associated with widespread ice stagnation and by 7,000 BP only small remnants of Laurentide ice remained.

Deglaciation of the Cordilleran ice sheet

Pattern of events

The timing of Cordilleran ice retreat is not known precisely although Mullineaux et al. (1965) have suggested that ice was still advancing into the Puget Lowland by 15,000 BP. Similarly, the Cordilleran ice sheet in northern Washington appears to have been retreating after the occurrence of the last major jökulhlaup of glacial Lake Missoula at 13,000 BP (Baker 1983a,b; Baker and Bunker 1985; Waitt 1985). Clague (1981) proposed that, by 11,000 BP, the Fraser Lowland was already free of ice that had probably receded to near present limits by 10,000 BP. He noted, however, that the pattern of deglaciation appears to have been interrupted in the Rocky Mountain trench region by a Lateglacial ice advance, possibly during the Younger Dryas (Clague 1975). Farther east, in the lee of the Rocky Mountains, Late Wisconsin precipitation starvation resulted in the development of an ice-free corridor between the Cordilleran and Laurentide ice sheets (Rutter 1984) that approximately extended southwards from the area of Jasper/ Edmonton to the present International Boundary (Rutter 1984). A second unglaciated enclave also appears to have developed in northern and western Yukon. On the Pacific coast, the independent ice caps that appear to have developed over Vancouver and Queen Charlotte Islands (Clague 1981) were subject to widespread deglaciation that was mostly completed by 13,700 BP (Clague et al. 1982).

During the deglaciation of Cordilleran ice in Montana, several extensive ice-dammed lakes were produced along the southern margin of the ice sheet. The largest of these ice-dammed lakes, Lake Missoula, was periodically produced as a result of ice advance and the creation of an ice barrier in the Clark Ford River valley near the Montana–Idaho border. Breaching of the ice barrier by lake

Plate 3 Luomusjarvi esker of Late Weichselian age, Utsjoki, northern Finland.
Photo M. Punkari.

Plate 4 Gorge eroded in rhythmically-bedded lake sediments of glacial Lake Missoula
(Burligame Canyon). The approximate position of the Mt. St Helens set-S volcanic ash
horizon is also shown (arrow). The ash layer, dated at approximately 13,000 BP is
overlain by eleven flood-deposited rhythmites and is capped by loess (after Waitt 1985).
Photo A. Werrity.

waters occurred on several occasions between 16,000 and 12,000 BP and was associated with the catastrophic drainage westwards (jökulhlaup) of 2,000 km³ of water (Figure 5.6A). These catastrophic floods are considered to have eroded the Channeled Scablands of the Columbia River Basin, Washington (Plate 4), as well as having deposited considerable volumes of sediment on the continental shelf and slope (Waitt 1980, 1985) (see Chapter 8).

THE ANTARCTIC AND GREENLAND ICE SHEETS DURING GLOBAL DEGLACIATION

The deglaciation chronology of Antarctic ice is largely unknown although it is generally believed that during the transition from the last glacial maximum to the Holocene, the East Antarctic ice sheet remained relatively stable while the West Antarctic ice sheet was subject to major changes (Stuiver *et al.* 1981). This is thought to have been dictated largely by the nature of ice shelf evolution in the Ross, Weddell and Amundsen Seas. The limit of the Ross Sea ice sheet in West Antarctica appears to have reached its maximum position between 21,200 and 17,000 BP, possibly due to grounding as a result of sea level lowering (Stuiver *et al.* 1981) (Figure 3.7). In general, the glacier fluctuations of the West Antarctic ice sheet appear to be broadly synchronous with those associated with the ice sheets in the northern hemisphere – in particular, the ending of full glacial conditions near 14,000 BP (Broecker and Denton 1990a,b). Heusser (1989) has also argued that Antarctica was subject to climatic warming soon after 15,000 BP while Clapperton and Sugden (1990) consider that in the Ross Embayment, Antarctica, ice recession after the last glacial maximum was underway slightly later, by 13,000 BP. Broecker and Denton (1990a,b) pointed out that the melting of Antarctic ice at this time took place during a period of accelerated worldwide deglaciation, despite the fact that the Milankovitch summer insolation signals in both hemispheres must have been out of phase (see Chapter 13).

Most of the Greenland ice sheet moraines that were produced during the last glacial maximum occur offshore and are not well dated. Controversy surrounds the postulated late Weichselian ice sheet limits for northern Greenland where one view is that extensive areas remained unglaciated due to extreme aridity while the opposing view is that of a complete northern Greenland ice cover that was confluent with ice over Ellesmere Island (Andersen 1981). Investigation of moraine sequences in East Greenland indicate the former occurrence of a major ice advance during the Younger Dryas that was of widespread extent and locally may have exceeded the inferred Late Weichselian ice sheet limit (Ten Brink and Weidick 1973; Funder and Hjort 1973). The Greenland ice cores show that a surface cooling of 6°C took place during the Younger Dryas (Dansgaard *et al.* 1989). By contrast, there is no signal in the Antarctic ice core record to show surface cooling during this period. However, Heusser (1989) has provided palynological data to suggest that the Antarctic continent may have been influenced by a episode of cooling between 13,500 BP and 10,000 BP.

THE ENIGMA OF YOUNGER DRYAS COOLING

A major ice advance appears to have occurred in NW Europe, the USSR and Spitsbergen (Kind 1972; Sissons 1979a; Mangerud *et al.* 1979; Boulton 1979) between 11,000 and 10,000 BP. This Younger Dryas ice advance in NW Europe has traditionally been interpreted as having resulted from precipitation variations due to latitudinal changes in the positions of the polar atmospheric and oceanic fronts in the North Atlantic (Ruddiman and McIntyre 1981a,b) (Figure 5.4). Implicit in this explanation is that since North Atlantic polar front fluctuations provide the forcing mechanism, the Younger Dryas stadial was a climatic deterioration that was confined to NW Europe (Mercer 1969).

Mott *et al.* (1986), Grant (1987, 1989) and Wright (1989) have suggested that a Younger Dryas ice advance may have occurred throughout Atlantic Canada. Evidence for an ice advance during the Younger Dryas has also been provided for part of the Cordilleran ice mass by Clague (1975) and Armstrong (1981) and for the nearby ice masses in the Cascade Ranges by Porter (1978). However, there can be little doubt that evidence for a Younger Dryas climatic deterioration is lacking for the southern and western sectors of the Laurentide ice sheet. Certainly, ice advances did take place in this region between 11,000 and 10,000 BP but these are mostly attributable to ice surges. Ironically, widespread

Plate 5 Large rounded boulders on floor of Roaring River channel in the Kaiashk system. This channel complex carried meltwater overflow from Lake Agassiz to the Nipigon Basin and Lake Superior during the Younger Dryas. Photo J. Teller.

retreat of ice in the Lake Superior region at this time is considered by Broecker *et al.* (1989) and Broecker and Denton (1990a,b) to have acted as a trigger mechanism that started a large-scale influx of meltwater into the North Atlantic that led to Younger Dryas cooling (see Chapter 8).

There is a growing body of evidence, however, from other areas of the world of ice advance during the Younger Dryas. This is of great significance since the widespread advance of mountain glaciers in the southern hemisphere during the Younger Dryas cannot easily be equated with meltwater influx into the North Atlantic! For example, Clapperton (1985) has proposed that a significant ice advance took place during this period in the Ecuadorian Andes, southern Peru (Clapperton and McEwan 1985; Mercer and Palacios 1977) and perhaps also throughout Patagonia. Similarly, Porter (1975) has argued that a major alpine readvance of glaciers took place in South Island, New Zealand, during the Younger Dryas. The evidence for global cooling during the Younger Dryas, outside Europe, is summarised by Rind *et al.* (1986). These authors developed a general circulation model to investigate whether or not colder sea surface temperatures in the North Atlantic and decreased air temperatures may have been sufficient to account for the observed patterns of cooling. Their results indicated good agreement between the results obtained by climate modelling and the palaeoclimatic evidence.

Thus the field evidence is contradictory. On the one hand, if the Younger Dryas of NW Europe resulted from the southward migration of the polar oceanic front in the North Atlantic, then a Younger Dryas ice advance should not have occurred elsewhere throughout the world. On the other hand, the occurrence of a Younger Dryas climatic deterioration throughout the northern hemisphere renders the model of North Atlantic polar front fluctuations an inappropriate forcing mechanism for the Younger Dryas event in NW Europe.

An attempt to reconcile these opposing views for the field evidence from Atlantic Canada has been made by Vernal and Hillaire-Marcel (1987) who suggested that the large-scale influx of Laurentide meltwater into the Labrador Sea prior to 11,000 BP may have resulted in a southward shift of the North Atlantic Drift and a consequent Younger Dryas cooling over NW Europe. These authors considered that frequent cyclone development took place over the Labrador Sea and the Atlantic Provinces throughout the Late Wisconsin and that abundant supply of precipitation to the eastern Laurentide ice sheet was provided by cyclones that tracked from SE to NW in this region. This view is a modification of that proposed by Mercer (1969) who argued that cooling of the North Atlantic surface waters took place during the Younger Dryas due to an influx of glacial meltwaters from the Arctic Ocean. In contrast to Vernal and Hillaire-Marcel (1987) and Mercer (1969), the model of Younger Dryas cooling proposed by Broecker and Denton (1990a,b) is different since the influx of meltwater into the North Atlantic is argued to have taken place almost instantaneously (see Chapter 8). However, all of these theories may prove to be incomplete explanations for the Younger Dryas if the cooling took place worldwide.

SUMMARY

Eurasia

The deglaciation history of the last ice sheets in the USSR, Scandinavia and the British Isles is known only in approximate outline. In European USSR and northern Scandinavia widespread ice thinning and retreat was underway by 16,000 BP and may have continued until between 13,500 and 13,000 BP. In European USSR and western Siberia, the pattern of ice-marginal oscillations was greatly influenced by the presence of ice-marginal lakes, into which ice masses may have surged periodically. It is generally believed that, in the British Isles, deglaciation was completed by 10,000 BP. However, the disappearance of ice from the USSR may not have been completed until after 9,000 BP while in Scandinavia the melting of the remaining ice masses may not have taken place until nearer 8,000–8,500 BP.

After 13,000 BP, there was rapid ice thinning and retreat in all areas while in the North Atlantic the polar oceanic and atmospheric fronts shifted northwards to positions similar to present. In Scandinavia, this period of interstadial warmth continued until near 11,000 BP although it was interrupted by a brief episode of cooling between 12,000 and 11,800 BP. In European USSR and central Siberia as well as Spitsbergen, a very pronounced ice sheet advance took place between 11,000 and 10,000 BP. This was paralleled by an ice sheet readvance of similar magnitude in Scandinavia as well as by the widespread build-up of ice in Scotland and the southward migration of the polar front in the North Atlantic. In southern Scandinavia, a very large ice-dammed lake (the Baltic Ice Lake) was formed between 12,800 and 10,450 BP. Its eventual drainage into the North Sea/North Atlantic region at 10,450 BP may have profoundly affected patterns of oceanic and atmospheric circulation.

North America

As in Eurasia, the melting history of the Laurentide and Cordilleran ice sheets is only known approximately. After the last glacial maximum, the western part of the Laurentide ice sheet remained relatively arid and was subject to early ice retreat (perhaps as soon as 17,000 BP). By contrast, the eastern section of the Laurentide ice sheet may have remained dynamically active until 14,000 BP owing to its nourishment by snow precipitation from the western Atlantic. Farther west, the Cordilleran ice sheet was subject to widespread thinning and retreat soon after 15,000 BP that resulted in the virtual disappearance of most ice by 10,000 BP.

During deglaciation along the southern margin of the Laurentide ice sheet, the non-synchronous advance and retreat of numerous ice lobes is attributable to the occurrence of several major ice surges during the overall retreat of the ice. Elsewhere, the drainage of ice-dammed lakes that had formed along the southern margin of the Cordilleran and Laurentide ice sheets resulted in catastrophic flooding. The largest floods that occurred prior to 13,000 BP took place westwards from glacial Lake Missoula and led to the development of the Channeled Scablands of the Columbia River Basin.

Between 13,000 and 11,000 BP there was a period of accelerated ice retreat and thinning that was marked not only by decay of ice in the Great Lakes area but also by widespread deglaciation throughout the Atlantic Provinces of Canada. Numerous ice advances took place during this time period but most, if not all, were due to ice surging. The dynamics of ice surging were largely influenced by the distribution of proglacial lakes. The most important of these was glacial Lake Agassiz that initially formed shortly after 12,000 BP. Large volumes of glacial meltwater were discharged into the Gulf of Mexico. Later diversion of lake drainage between 11,000 and 10,000 BP, due to ice retreat, led to the discharge of Agassiz meltwaters into the western North Atlantic.

Although there is pollen evidence for cooling at 11,000–10,000 BP in Maritime Canada, there is limited evidence from Quebec and Newfoundland that the Laurentide ice sheet was subject to a major ice advance during the Younger Dryas. The occurrence of a climatically induced ice advance at this time is, however, complicated by the possibility that certain ice advances may have been due to ice surging. One very large surge that took place in the Lake Superior area between 9,900 and 9,500 BP led to a temporary (circa 500 year) end of Lake Agassiz overflow into the North Atlantic. Progressive ice thinning and retreat eventually led at 8,000 BP to the final disintegration of the Laurentide ice sheet over Hudson Bay when drainage of Lake Agassiz waters took place through the Hudson Strait into the western Atlantic. By 7,000 BP only small remnants of the Laurentide ice sheet remained.

Not much is known about the history of Antarctic and Greenland ice sheets during the time when northern hemisphere ice sheets were melting. It is generally believed that the East and West Antarctic ice sheets decreased in size but it is not known by how much and at what rate. The most recent view is that climatic warming may have commenced soon after 15,000 BP while there may have been an episode of cold conditions between 13,500 and 10,000 BP (Heusser 1989). A similar controversy surrounds the Lateglacial evolution of the Greenland ice sheet. Here, the melting history is additionally complicated by the possibility that in eastern Greenland, ice advance during the Younger Dryas, as in Spitsbergen, may have been more extensive than during the last glacial maximum.

Global effects

A major ice advance appears to have taken place during the Younger Dryas between 11,000 and 10,000 BP. Recent analysis of the rates of glacio-eustatic sea level rise that occurred during this period of ice sheet melting indicate that the rate of sea level rise slowed down markedly during the Younger Dryas – a view in agreement with the glacial evidence (Fairbanks 1989). Equally, the periods of rapid ice melting between 13,000 and 11,000 BP and between 10,000 and 7,000 BP are characterised by very rapid rates of sea level rise (Fairbanks 1989). Notwithstanding the importance of global sea level rise throughout the period of ice melting, the climatic consequences of several major floods due to the drainage of ice-dammed lakes appear also to have been considerable. Indeed, Broecker *et al.* (1989) have argued that the emptying of Lake Agassiz floodwaters into the North Atlantic between 11,000 and 10,000 BP may actually have triggered global cooling (see Chapter 8). The largest catastrophic floods, however, occurred much later at 8,000 BP. These were due to the final disintegration of the Laurentide ice sheet over Hudson Bay and led to an almost instantaneous rise in global sea level of between 0.2 and 0.4 m. The environmental changes that took place during the melting of the last ice sheets led to immense changes in the redistribution of mass across the surface of the Earth. At present we know almost nothing about the way that the Earth's crust and interior responded to these changes.

NOTES

1   This calculation is based on (1) a cross-channel area for the Hudson Strait of $30\,km^2$ ($120\,km \times 0.25\,km$) at 8,000 BP, (2) an average flood velocity of $20\,ms^{-1}$, (3) a lake volume of $100,000\,km^3$ and (4) a global ocean area of $361 \times 10^6\,km^2$.

6

ICE AGE PERIGLACIAL
ENVIRONMENTS

INTRODUCTION

Periglacial environments are those characterised by non-glacial processes and features of cold climates on land affected by frost action, regardless of their proximity to glaciers (Washburn 1979; Pewe 1983). Most periglacial regions are underlain by perennially frozen ground or permafrost (Pewe 1983). Permafrost regions are usually defined as those areas of the ground in which a temperature below freezing has existed continually for a long time (from two years to tens of thousands of years) (Muller 1947). The growth of permafrost is dependent upon the radiative heat loss from the ground during winter being greater than the supply of heat to the ground surface during the summer months. Conversely, permafrost decay may occur when summer heating is greater than winter heat loss. At present, periglacial environments occur over 20 per cent of the Earth's land surface although during the last glacial maximum an additional 20 per cent may have been affected (French 1976).

The aim of this chapter is to use evidence of former periglacial activity to reconstruct spatial and temporal variations in climate for the Late Quaternary. Relict features indicative of former permafrost are of particular value in the reconstruction of past environments since they provide valuable information on former ground temperature conditions (Washburn 1979; Pewe 1983). Relict periglacial features that are not diagnostic of former permafrost conditions are of less value in palaeoenvironmental reconstruction since it is often very difficult to attribute particular phenomena to former climatic conditions. Accordingly, in the following sections, most consideration is given to the evolution of permafrost environments during the Late Quaternary, particularly with regard to the USSR and North America for which most data are available (see pp. 114–22). The distribution and palaeoclimatic significance of Late Quaternary loess is considered in Chapter 9.

PROBLEMS IN PERMAFROST RECONSTRUCTION

The identification of relict features that are indicative of former permafrost is, of necessity, dependent on the recognition of analogous features that are presently being formed in permafrost regions. The most common features that indicate the former presence of permafrost are fossil rock glaciers, pingo scars, cryoplanation terraces and ice wedge casts. Of these, only permafrost wedge structures are common in sediments predating the last glacial maximum. Accordingly, since it is not possible to date ice wedge casts directly, most reconstructions of Late Quaternary permafrost history have been largely based on radiometric dating of organic sediments that have been disturbed by episodes of ice wedge growth.

There are major difficulties in such types of reconstruction since environmental characteristics of present-day permafrost phenomena in high latitudes cannot be compared easily with former features indicative of permafrost that were produced in the middle latitudes during the last glaciation. This difference largely reflects the patterns of solar radiation that affected the permafrost landscapes located south of the last major ice sheets in the northern hemisphere. In these environments, the relatively high number of freeze–thaw cycles per year may have led to a much greater intensity of geomorphic processes than those presently characteristic of high latitude permafrost environments (French 1976).

Another difficulty is that since the characteristics of the Earth's orbit around the Sun have changed through time, there have been major long-term changes in the duration and intensity of winter freezing and summer thaw (Chapter 13). For example, 11,000 years ago the Earth was closest to the Sun during the northern hemisphere summer and farthest from it during winter. Consequently, the warmer summers and colder winters at this time may have led to the development of Lateglacial periglacial phenomena that were quite different in terms of their palaeoclimatic significance from those of the present. An additional problem is that permafrost leaves no direct trace of its former presence once it has thawed. The geomorphological and stratigraphical traces are indirect and are often difficult to interpret.

There are several landforms that, in relict form, have been used as indicators of former permafrost and have thus provided information on former temperature conditions. However, there is a great deal of misunderstanding about whether the indicators of former permafrost provide information on former ground or air temperatures. As a general guide, the mean annual ground temperature is between 2 and 6°C warmer than the mean annual air temperature (Pewe 1983). Conversion from mean annual ground temperature to mean annual air temperature is a hazardous exercise and is significantly affected by the presence or absence of a winter snow cover and the type of sediment in which the features are developed (Pissart 1987).

Polygonal cracks associated with the casts of former ice wedges may be indicative of former mean annual ground temperatures between −2°C and −10°C (Washburn 1979) (Plate 6). Ice wedge cracks, however, may require winter ground surface temperatures of −15°C or below (Pewe 1983). Pewe (1975) maintained that the mean annual air temperature of areas in which ice wedges are presently being formed is between −6 and −8°C.

Closed system pingos are considered indicative of a mean annual ground

Plate 6 Stratigraphically superimposed ice wedge casts of Late Quaternary age at Dix's Pit, Stanton Harcourt, Oxfordshire, England. Photo M. Seddon.

temperature of less than 0°C and maximum mean annual air temperatures of −6°C (Washburn 1979). By contrast, open system pingos occur in areas where the mean annual air temperature is less than −2°C (Washburn 1979). Additional features that are indicative of former permafrost, and hence mean annual temperatures of less than 0°C, include inactive lobate rock glaciers and cryoplanation terraces. There are numerous other features that provide more equivocal evidence of former permafrost conditions. These include various types of patterned ground and mass movement phenomena (e.g. fossil sorted and non-sorted circles and polygons, gelifluction lobes and blockfields). Fossil dunes and loess are different types of palaeoenvironmental indicators since although they do not provide data on former temperatures, they provide valuable information on patterns of former atmospheric circulation. These are discussed more fully in Chapter 9.

PERMAFROST EVOLUTION

The evolution of permafrost in response to climatic warming and cooling is relatively slow. For example, in northeast Siberia permafrost has been continuously present since the Early Quaternary (Washburn 1979). During the Late Quaternary, changes in permafrost thickness have taken place at a considerably slower rate than the growth and decay of ice sheets and sea level changes. For example, extensive areas of relict permafrost occur across the sea floor of the Arctic continental shelf due to the rise in sea level during the Holocene (Pewe 1983).

In those areas where permafrost is present today (e.g. Siberia and Alaska), stratigraphic studies have enabled the identification of generations of permafrost-related ice wedge casts. In the USSR, dating of palaeosols and loessic sediments has enabled a crude permafrost history of the Late Quaternary to be established. In other areas where permafrost is no longer present, the identification of landforms and sedimentary structures indicative of former permafrost has similarly enabled reconstructions of air and ground temperature fluctuations for this period.

In theory, the development of ice sheets in Eurasia and North America during the Late Quaternary should have impeded the growth of permafrost over those areas upon which they developed. This view is based on the premise that the ice sheets insulated the underlying ground surface due to the trapping of geothermal heat. However, there is only a poor relationship between the distribution of the last ice sheets in the northern hemisphere and the present distribution of permafrost. It is possible, for example, that during the build-up of the last ice sheets, ice advanced over areas already underlain by permafrost. A more informative comparison is between the distribution of areas of thickest permafrost during the last glaciation and the limits of the last ice sheets (Figure 6.1). For example, in the Soviet Union, most areas where permafrost attained thicknesses of greater than 800 m occur outside the limits of the last (Valdai/Sartan) ice sheet (Baulin and Danilova 1984). However, such great thicknesses of permafrost are not

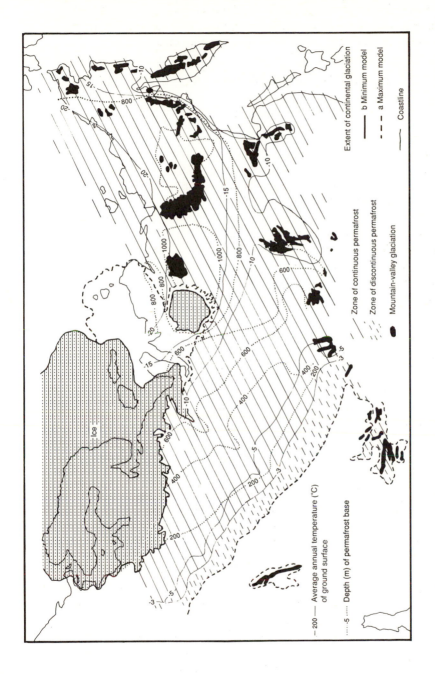

Figure 6.1 Map of permafrost distribution during the Valdai (Sartan) glaciation (after Baulin and Danilova 1984).

considered the sole product of the last glacial period. Important exceptions to this pattern are those areas in which ice-dammed lakes developed during the last glaciation, since in these areas permafrost growth was prevented due to the insulation of the ground provided by the overlying water masses.

LATE QUATERNARY PERMAFROST IN THE USSR

Introduction

The widespread degradation of permafrost throughout the USSR during the last interglacial (isotope substage 5e) was followed by the growth of thick permafrost during the Late Quaternary. Indeed, in unglaciated areas around the Arctic Circle, permafrost growth during this period may have reached thicknesses of up to 700 m (Baulin and Danilova 1984). During the last glacial maximum, continuous permafrost existed almost as far south as 50°N while areas of discontinuous permafrost existed even farther south, perhaps as far as 48/49°N (i.e. 400–600 km south of the ice sheet margin) (Figure 6.1). However, the evolution of permafrost during the Late Quaternary is complex with several periods of permafrost degradation having alternated with periods of permafrost growth during glacial periods.

Permafrost development prior to the Late Valdai/Sartan glaciation

Studies of Late Quaternary permafrost history in the USSR have provided valuable information on the patterns of environmental change between the last interglacial and the last glacial maximum. In places there is evidence of different generations of permafrost features indicative of former changes in permafrost development. Velichko and Nechayev (1984) consider that the Late Quaternary was affected by three distinct periods of periglaciation, most clearly defined in European USSR and least pronounced in eastern Siberia (Figure 6.2).

Two distinct periods of permafrost growth have been recognised for this period. The first cold period appears to predate 70,000 BP and was characterised by sporadic permafrost development and the formation of ice wedge polygons, involutions and cryoturbation structures. A later period of climatic cooling and permafrost growth culminated between 60,000 and 50,000 BP (Figure 6.2). During this time period, soils of last interglacial age were subject to considerable cryoturbation while elsewhere ice wedge polygons were produced. Velichko and Nechayev (1984) consider that this period of periglaciation represents a pronounced expansion of permafrost areas and a significant lowering of mean annual ground temperatures. They consider that near the end of this period, the southern limit of continuous permafrost may have been displaced 6 to 8° of latitude south of its present position, reaching the centre of the Russian Plain where mean annual ground temperatures may have been as low as −3°C. Thereafter, a brief interlude of climatic warming preceded major permafrost expansion during

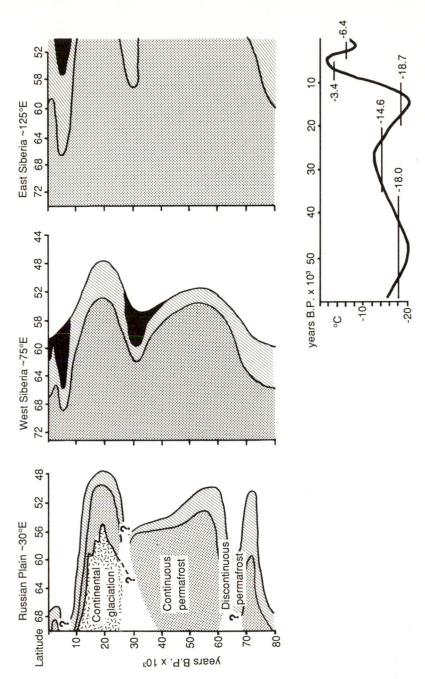

Figure 6.2 Dynamics of Late Quaternary permafrost in the USSR showing marked zonal variations. A profile of permafrost temperatures (curved line) for this period is also shown (the average values are shown by straight lines) (after Baulin and Danilova 1984).

the Late Valdai/Sartan glaciation. The timing of this period of interstadial warmth has been estimated between 29,000 and 24,000 BP by the dating of fossil soils (palaeosols) (Velichko and Nechayev 1984). The latter authors stress that in the Russian Plain this interstadial was dominated by very cold conditions and can only be considered as 'warm' in relation to the extreme cold that characterised the rest of the Valdai interval.

The regional variations in permafrost evolution in the USSR during the Late Quaternary show some interesting trends (Figure 6.2). During this period, European USSR appears to have been characterised by pronounced fluctuations in permafrost growth and degradation. By contrast, eastern Siberia was subject to significantly less marked changes in permafrost distribution and thickness. This west to east variation highlights the important influence of continentality on permafrost evolution in the USSR.

Permafrost development during the Late Valdai/Sartan glaciation

A very extensive period of periglacial activity took place after 24,000 BP at the beginning of the Late Valdai glaciation. During this period, there was considerable disturbance of soils that had formed during the preceding interstadial (29,000–24,000 BP). In general, the distribution of continuous permafrost estimated for this period does not bear any spatial relationship to the distribution of Late Valdai ice sheets.

The regional differences in the nature of permafrost growth in different parts of Eurasia during the Late Valdai were considerable. Thus, in northeast Siberia where present permafrost thicknesses normally exceed 800 m (locally reaching up to 1,400 m), the increase in permafrost thickness during the last glacial period may only have been between 100 and 250 m (Baulin and Danilova 1984). By contrast, permafrost in European USSR, where it is absent today, may have reached thicknesses of up to 200 m during the Late Valdai. The reconstruction of permafrost conditions in the USSR for the last glacial maximum shows clearly that the most severe conditions existed in eastern and northeastern Siberia. In these areas, average annual ground temperatures may have fallen below −15°C (Figure 6.2).

The southern limit of continuous permafrost in the USSR for the Late Valdai has largely been defined by the southern limit of sets of ice wedge casts (pseudomorphs). These show that the development of ice wedge polygonal networks was very widespread during this period. In many places, ice wedges were formed in loess sediments although they were also produced in fluvioglacial sediments and glacial diamicton. Baulin and Danilova (1984) consider that it is possible to establish a relationship between the spacing of ice wedge polygon cracks and mean annual ground and air temperatures for the Late Valdai. They suggested that the ice wedge polygons that formed during this period are more closely spaced than those forming at present while individual wedges were thicker (Wright and Barnowsky 1984). By contrast, the ice wedge polygons that were produced in

European USSR were smaller than those that formed in Siberia. In general the ice veins extended to depths between 2.5 and 5 m while the vein ice volume may only have been 8–12 per cent of the mass – much less than for features presently forming in Siberia (Wright and Barnowski 1984). Baulin and Danilova (1984) concluded that most of European USSR was characterised by average annual ground temperatures of between −3 and −10°C (Figure 6.1).

Late Valdai permafrost climates

During the Late Valdai glaciation, the presence of the Eurasian ice sheet enabled the formation of a permanent area of high pressure across European USSR. In the unglaciated areas of western and eastern Siberia and European USSR, the seasonal development of low pressure enabled the formation of a 'summer glacial monsoon' (Velichko 1984). In this type of circulation, of which there is no analogue today, the unglaciated areas were invaded by dry air masses from neighbouring ice sheets.

During winter, the lack of snow precipitation resulted in an increased susceptibility of the ground to heat loss. This widespread surface cooling led not only to the creation of high pressure across the entire Eurasian continent but also resulted in widespread permafrost aggradation. It is likely that anticyclones developed over (a) the Eurasian ice sheet, (b) eastern Siberia (centred over the Verkhoyansk ice sheet) and (c) over the Tibetan Plateau. The establishment of these high pressure cells was additionally favoured by an Arctic Ocean sea ice cover, increased continentality in Arctic USSR due to sea level lowering, and changes in atmospheric and oceanic circulation over the North Atlantic. This latter aspect is crucial since the southward displacement of the polar atmospheric and oceanic fronts in the North Atlantic prevented the invasion of cyclonic air masses into European USSR and hence resulted in increased aridity.

An additional, and as yet unknown, factor of great importance in understanding atmospheric circulation during the Late Valdai in Eurasia is the influence exerted by the mid latitude jet stream. It is likely that jet stream flow was split around the Eurasian ice sheet with streams flowing westwards both to the north and south of the ice sheet (Kutzbach and Wright 1985). Under such circumstances, jet stream convergence is likely to have occurred over eastern Siberia and may have played an important part in anchoring the position of air masses over eastern Asia and thus maintaining the aridity of the region.

It is almost impossible to reconstruct with any accuracy patterns of former atmospheric circulation as they developed during the Early and Middle Valdai. In particular the period of permafrost degradation after 30,000 BP is difficult to explain. Almost certainly, polar front fluctuations over the North Atlantic played a major role in altering the supply of precipitation to European USSR. However, the patterns of atmospheric circulation and the associated temperature and precipitation changes over European USSR are virtually unknown.

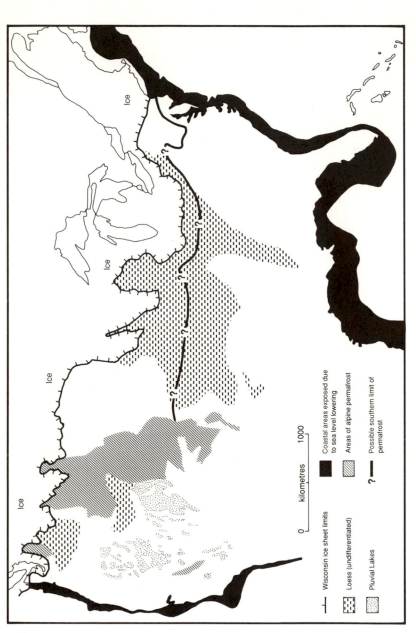

Figure 6.3 Palaeogeographic reconstruction of the United States for the last glacial maximum showing the extent of permafrost, loess and 'pluvial' lakes (after Washburn 1979; Pewe 1983).

LATE QUATERNARY PERMAFROST IN NORTH AMERICA

The continental interior of the USA

Most of the Late Quaternary relict permafrost phenomena that have, so far, been identified in North America were produced during the Late Wisconsin glacial maximum (Figure 6.3). There is relatively little information on permafrost environments prior to the Late Wisconsin. During the latter period, a permafrost environment existed south of the Laurentide and Cordilleran ice sheets as well as near other mountain glaciers and small ice cap complexes. Late Wisconsin permafrost was also widespread throughout all of unglaciated Alaska as well as in northwestern Canada (Pewe 1983).

In the United States most ice wedge casts of Late Wisconsin age occur within 100 km of the Laurentide ice sheet limit. They are particularly common in North Dakota as well as in the driftless area of Wisconsin and in New Jersey (Pewe 1983). The distribution of ice wedge casts in Wisconsin led Black (1969) to conclude that Late Wisconsin mean annual temperatures in this area were between −5 and −10°C. Barry (1983) has noted that these areas would have been particularly favourable for the growth of ice wedges since they are likely to have been characterised by light snowfall and strong winds. Fossil pingos have been described from north-central Illinois (Flemal 1976) while numerous inactive rock glaciers presumed to be of Late Wisconsin age have been described from south-central New Mexico (Blagbrough and Farkas 1968).

Pewe (1983) depicts the area of continuous and discontinuous permafrost in the United States as grading westwards into a widespread area of alpine permafrost that existed in northern Idaho, western Montana, Wyoming and Colorado. In these areas, alpine permafrost covered about 500,000 km^2 while the altitudinal limit of permafrost was approximately 1,000 m lower than it is today (Figure 6.3). Similarly, Galloway (1970) concluded that in the mountains of southwestern United States, relict slope deposits indicated that the timber line had been lowered by 1,300–1,400 m during the Late Wisconsin with mean annual temperatures having been 10–11°C lower than present. Pewe (1983) has estimated, for example, that mean annual air temperatures in Wyoming during the Late Wisconsin may have been between 9 and 11°C lower than present.

Alaska and northwestern Canada

At present, permafrost underlies 82 per cent of Alaska as well as all of northwestern Canada (Pewe 1983). Most areas of central and southern Alaska occur in the zone of discontinuous permafrost. In northern Alaska, however, particularly on the North Slope, continuous permafrost is widespread and exceeds 400 m in thickness near Barrow and Prudhoe Bay (Walker 1973). During the Late Wisconsin, glacio-eustatic sea level lowering resulted in the development of a vast unglaciated enclave (Beringia) that extended from northeastern Siberia to the margins of the Cordilleran and Laurentide ice sheets to the south and east (Hopkins 1972). The unglaciated landscape that was produced in this area was characterised by continuous permafrost. Ice wedges that formed during this period have been considered by Pewe (1975) to indicate the former occurrence of mean annual air temperatures of at least −6°C (Figure 6.4).

Cryoplanation terraces are the most widespread periglacial form of Late Wisconsin age in Alaska and northwestern Canada where they were active over an area of about 250,000 km². These have been interpreted by Reger and Pewe (1976) as indicative of former mean summer air temperatures between 2 and 6°C and mean annual air temperatures of about −12°C. However, many of the cryoplanation terraces are not well dated to the Late Wisconsin while the nature of the permafrost regime associated with their development is not yet known. Perm-

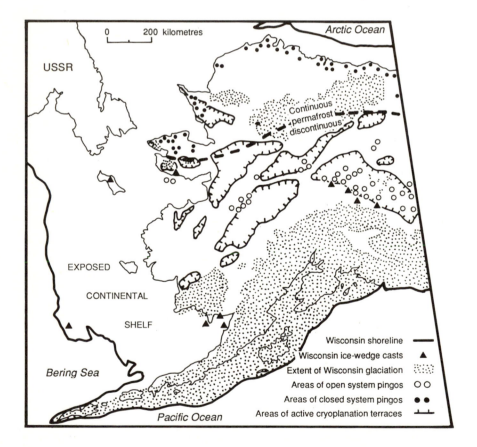

Figure 6.4 Periglacial environments in Alaska during the last glacial maximum (after Pewe 1983).

afrost is also considered to have developed on the continental shelves of Beringia that were exposed due to sea level lowering (Hopkins 1972) (Figure 6.4). In these areas relict permafrost occurs below present sea level, due to the postglacial rise in sea level.

In many parts of Alaska and northwestern Canada, the presence of pingos provides unequivocal evidence of permafrost (Figure 6.4). However, all of the pingos that have been identified appear to be Holocene in age. Pingos must have been actively forming during the Late Wisconsin, yet no record has, to date, been found for their existence (Pewe 1983). Yet the distribution of Alaskan pingos is intriguing since most occur in areas that remained unglaciated during the Late Wisconsin. An exception to this pattern is in the MacKenzie delta region where many pingos occur within the limits of Late Wisconsin ice. Most of the closed system pingos along the North Slope of Alaska occur in areas of continuous permafrost where the mean annual air temperature is below −5°C. Most open system pingos, however, occur in central Alaska in areas of discontinuous permafrost and usually in areas where the mean annual temperature is −2°C or lower (Figure 6.4).

Palaeoclimatic implications

The occurrence of a relatively narrow belt of permafrost south of the Laurentide ice sheet during the Late Wisconsin contrasts markedly with the very extensive areas of permafrost that developed south of the Eurasian ice sheet (compare Figures 6.1 and 6.3). French (1976) has argued that this may be partly due to the occurrence of the Laurentide ice sheet at much lower latitudes than the Eurasian ice sheet.

The evidence for permafrost conditions during the Late Wisconsin suggests that the mean annual air temperatures were lower than −6°C across a broad zone extending from Montana to New Jersey. In the areas of alpine permafrost, mean annual temperatures may have locally been as low as −15°C (Pewe 1983; Barry 1983). These estimates are consistent with numerical models of atmospheric circulation in North America for this period (e.g. Kutzbach and Wright 1985). The models indicate that the westerly jet stream may have split to both the north and the south of the Laurentide and Cordilleran ice sheets. In this way the zonal (west–east) winds would have have been anchored and increased in strength. Kutzbach and Guetter (1986) have shown that the southern jet would have favoured increased precipitation in the right entrance sector of the jet (over southwestern USA) and the left exit sector (in northeastern USA and eastern Canada). By contrast, aridity and low temperatures would have been characteristic of the intervening area. The combined effect of these processes was that widespread permafrost development did not take place south of 35⁰ latitude.

In Alaska, the widespread thickening of permafrost appears to have been closely related to the northern branch of the split jet stream. In general, it appears that central and northern Alaska as well as northwestern Canada were affected both by low temperatures and low precipitation, the latter enhanced by the topographic influence of the Cordilleran ice masses that presented an orographic barrier to precipitation from the Gulf of Alaska and the North Pacific. Large areas of Alaska would have occurred beneath the left entrance sector of the jet that, in turn, would have favoured lower tropospheric subsidence (i.e. high pressure) and reduced precipitation.

LATE QUATERNARY PERIGLACIAL ENVIRONMENTS IN EUROPE

In Europe, fossil ice wedge polygons, pingos and cryoturbation structures of Late Weichselian age are widespread. In addition there are extensive accumulations of dunes and loess. In certain areas Late Quaternary periglacial phenomena predating the Late Weichselian have been identified and these have been used to reconstruct longer term environmental changes. Several criteria have been used in Europe to reconstruct permafrost characteristics. For example, Maarleveld (1976) used inferred maximum mean annual air temperatures of −2°C for frost cracks and −6°C for ice wedge pseudomorphs to map the southern limit of permafrost (Figure 6.5). Similarly, the poleward limit of forest cover based on palynological criteria has been used to define the summer 10°C isotherm (Poser, cited in Washburn 1979). Although differences of opinion exist regarding the palaeoenvironmental significance of these indicators, the values provide some approximation to former environmental conditions.

Haesaerts (1984) has deduced a complex pattern of palaeoclimatic changes from stratigraphical studies of ice wedge pseudomorphs in the loess of Belgium (Figure 6.6). The climatic changes correspond well with those inferred from the Grand Pile pollen diagram (Mook and Woillard 1982) and show early periods of permafrost development at the isotope substage 5b cooling 'spike' as well at the isotope stage 5/4 boundary (Figure 6.6). Several later periods of permafrost development occurred in continental northwest Europe between 50,000 and 25,000 BP and were separated by periods of interstadial warmth (Karte 1987; Pissart 1987). These included the Moershoofd (between 50,000 and 45,000 BP), Hengelo (between 39,000 and 37,000 BP) and Denekamp interstadials (between 32,000 and 29,000 BP) (Pissart 1987) (Figure 6.6).

The coldest period of the last glaciation appears to have occurred during ice sheet build-up between 25,000 and 20,000 BP. Frenzel (1973) argued, using several criteria, that the lowering of mean annual air temperature during the last glacial maximum was at least 12°C in southern England, 15–16°C in southern France and near 13°C in eastern Europe. Similarly, Williams (1975) concluded that mean annual air temperature during the last glacial maximum in southern and central England was probably between −8 and −10°C. Together their data indicate very substantial temperature lowering in Europe during the last glacial maximum. They show quite clearly that most of France, Germany, Poland, Belgium, the Netherlands and southern Britain were characterised by permafrost

122

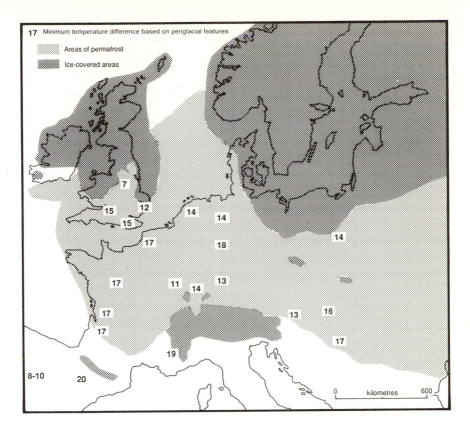

Figure 6.5 Permafrost in Europe during the last glacial maximum. The shaded and semi-shaded areas respectively indicate areas of ice cover and permafrost. The numbers indicate the estimated minimum temperature difference between the last glacial maximum and the present day (based on Washburn 1979).

and increased continentality during the last glaciation. Karte (1987) has estimated that the last glaciation in north-central Europe was associated with permafrost thicknesses of more than 100 m in the east and northeast but nearer 10 m in the west and southwest.

PALAEOCLIMATIC INFERENCES

During the last glacial maximum, the pattern of atmospheric circulation in Europe was overwhelmingly dominated by (a) the southward extension of the polar atmospheric front in the North Atlantic, (b) permanent high pressure over Scandinavia, the North Sea and the British Isles due to the presence of ice sheets, and (c) a local ridge of high pressure over an ice cap in the Alps. These three factors caused the westerly mid latitude jet stream to be displaced southwards to

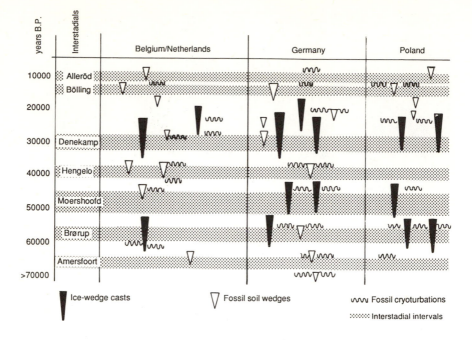

Figure 6.6 Stratigraphy of Weichselian periglacial structures in central Europe showing the complex nature of Late Quaternary climate changes (based on Haesaerts 1984 cited in Karte 1987).

flow from west to east across the Mediterranean. Thus, whereas the latter area was affected by the passage of numerous precipitation-bearing cyclones tracking eastwards, central and western Europe was dominated by cold and dry easterly winds generated along the southern margin of the ice sheets in Scandinavia and European USSR. The southward extension of permafrost and the considerable temperature lowering during this period (in many areas as far as 600 km south of the ice sheet border) may be attributable to this cause. The maritime influence of the North Atlantic on western Europe was scarcely felt owing to increased continentality caused by (a) the southward position of the polar oceanic and atmospheric fronts in the North Atlantic and (b) glacio-eustatic sea-level lowering.

SUMMARY

The evolution of permafrost that took place during the Late Quaternary was profoundly influenced not only by changes in temperature but also by changes in atmospheric circulation. These, in turn, were greatly influenced by the build-up and decay of immense continental ice sheets in the middle latitudes. A major feature of atmospheric circulation in Europe and North America during glacial periods was an enhanced zonal wind flow associated with the anchoring of a mid latitude jet stream south of the major ice sheets. The easterly winds that were generated south of the ice sheet margins, acting in conjunction with katabatic winds and low air temperatures, led to a negative heat budget at the ground surface and the consequent aggradation of permafrost.

Considerable progress has been made in reconstructing the extent of permafrost in Europe, Asia and North America for the Late Quaternary. The reconstructions that have been made for the Soviet Union are the most extensive. They show that prior to the last glacial maximum, permafrost growth was interrupted on two occasions. One of these periods of relative warming occurred between 70,000 and 60,000 BP and was most marked in European USSR. The second, and younger, period of warming appears to have taken place between 29,000 and 24,000 BP and was thereafter followed by the growth of the Late Valdai/Sartan ice sheet. Estimates of average ground temperature appear to suggest that it reached values as low as $-18°C$ during the most severe glacial episodes. During the last glaciation, vast areas of Asia and Europe were subject to the growth of permafrost. In North America, permafrost development was also widespread but was mostly confined to a narrow 80–200 km belt south of the Laurentide ice sheet. The marked temperature lowering also led to the widespread growth of alpine permafrost in parts of Idaho, North Dakota, Wyoming and Colorado.

A detailed reconstruction of Late Quaternary environmental changes has been derived from studies of periglacial phenomena developed in the Belgian loess (Haesaerts 1984). These show broad agreement with the timing of the climatic changes identified for the Soviet Union. However, additional climatic fluctuations are indicated; in particular cooling episodes during isotope substage 5b and the isotope stage 5/4 cooling phase as well as several oscillations during isotope stage 3.

7

LAKES, BOGS AND MIRES

INTRODUCTION

During the last ice age, there were dramatic changes in the distribution of the world's lakes. In many non-glaciated environments, changing patterns of precipitation, evaporation and temperature led to major fluctuations in the size of many lakes. In addition, the development of the major northern hemisphere ice sheets resulted in the creation of extensive ice-dammed lakes (see Chapter 5). For a long time it was believed that many lakes in non-glaciated regions fluctuated in response to the growth and decay of the Laurentide and Cordilleran ice sheets. It was thought that high, so-called, 'pluvial' lake levels occurred during periods of glaciation, as a result of the combined influence of lower temperatures, increased precipitation and decreased evaporation. However, this interpretation is now considered to be mistaken and, instead, it is thought that lakes have responded to Late Quaternary global climate changes in very complex ways.

The lakes that developed in non-glaciated areas during the Late Quaternary provide quite detailed evidence of past environmental changes since the sediments that have accumulated on their floors provide undisturbed records of lake sedimentation over relatively long time periods. By contrast, the evidence for former ice-dammed lakes provides information on relatively short-lived palaeoenvironmental changes restricted to the duration of the last glaciation. The aims of this chapter are threefold.

1   To describe and explain long-term patterns of climate change inferred from long sediment sequences in selected lake basins, bogs and mires.
2   To provide an account of Late Quaternary lake level changes in non-glaciated areas.
3   To consider why fluctuations in lake level took place in particular regions at specific times during the Late Quaternary.

THE LONG-TERM EVOLUTION OF LAKES, BOGS AND MIRES

The siltation of lakes, coupled with the expansion of the areas of plant growth around the edges of lakes, eventually leads to the disappearance of all areas of open water. All waterlogged areas where peat is formed due to reduced vegetal decay under anaerobic conditions are known as mires. The succession from open water lake conditions to mire, and eventually bog, is referred to as a hydrosere and is usually accompanied by a change in the sediments from muds to peats (Lowe and Walker 1984).

In many lake, bog and mire sediment accumulations, a record is provided of the changing processes that have affected the slopes of individual catchments. Many aspects of the basin sediment stratigraphy can be investigated in order to reconstruct former changing environmental conditions. The greatest problem is often to select those attributes of the sediments that are most likely to provide valuable insights into former environmental changes.

For most sediment accumulations, there are often marked changes in pollen content and sediment lithostratigraphy. For example, in many mid latitude lake basins, former periglacial conditions in the catchment are often associated with increased rates of minerogenic sediment deposition into lakes and a corresponding decrease in arboreal pollen influx due to a greatly reduced vegetation cover. By contrast, more temperate conditions are associated with organic-rich lake sediments and an increased vegetation cover. Detailed information on Late Quaternary environmental changes has been obtained from studies of the ratio of arboreal pollen to that of grasses and herbs. Patterns of former environmental change can also be established by studies of sediment geochemistry (e.g. palaeo-salinity studies) as well by using various palaeomagnetic indicators (e.g. magnetic susceptibility, inclination and declination). Frequently, studies of the diatom flora within lake sediments provide important information about former environmental changes, as well as lake level changes.

In lakes whose levels have fluctuated significantly throughout the Late Quaternary, additional types of palaeoenvironmental information are available. For example, formerly low lake levels are indicated by stratigraphic unconformities while complete drying of the lake floor is often indicated by the presence of evaporites (accumulations of salts). Periods of formerly high lake level are often represented by lake strandlines. However, the response of any lake to climate change is often complex and not only depends on its hydrology but is also influenced by extrinsic factors such as variations in precipitation and tectonic activity. Lake level variations are additionally influenced by whether the lake basins occur in closed or open systems. In general, open lake basins are those that possess river outflows or are those characterised by intrinsic influences, for example groundwater leakage. By contrast, lake level changes in closed lake basin systems, in which groundwater losses are negligible, may be more sensitive indicators of past climate changes.

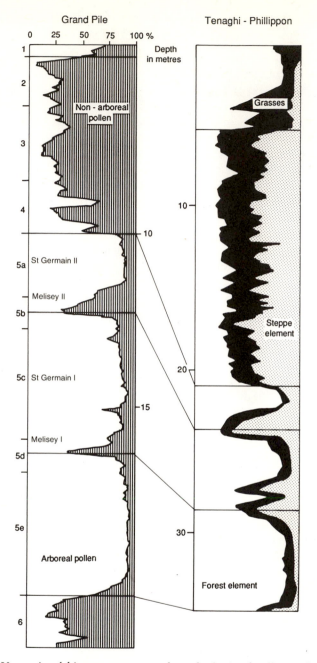

Figure 7.1 Vegetational history reconstructed on the basis of pollen analysis at Grand Pile, France (left) and at Tenaghi–Phillippon, Greece (right) (based on Van der Hammen *et al.* 1971; Mook and Woillard 1982; Eronen and Olander 1990). The oxygen isotope stratigraphy is shown (left).

LONG SEDIMENT SEQUENCES OBTAINED FROM LAKES, BOGS AND MIRES

Introduction

Studies of lake sediment stratigraphy have provided a record of environmental change that extends through the entire Quaternary (e.g. at Sabana de Bogota in Colombia – Van der Hammen (1974)). In most places, however, the record of environmental change provided by lake sediments covers much shorter time periods. There are relatively few sediment sequences from lakes, bogs and mires that span the Late Quaternary. Those that are known therefore provide very valuable information on patterns of former climatic changes. Caution must, however, be exercised in the interpretation of records of past vegetational changes since the changes that are indicated need not mirror past climatic changes. This is because changes in vegetation respond to hydrosere change and invariably lag behind changes in climate.

In the following sections, four examples are used to illustrate the types of palaeoclimatological information for the Late Quaternary that have been provided through the study of lake, bog and mire sediment sequences. These include three examples of long records of Late Quaternary vegetational history derived from bogs and mires (Grand Pile, Les Echets and Tenaghi-Phillippon) as well as one based on the use of lake sediments (Lake George, Australia).

Grand Pile: France

A detailed pollen record from the Grand Pile peat bog in the Southern Vosges Mountains of France shows a complex series of vegetational changes that span the last 140,000 years (Woillard 1978) (Figure 7.1). Near the base of the core, the vegetation is characteristic of the last (Eemian) interglacial. Farther up the core, vegetation conditions indicative of temperate environments (St Germain I and II) are interrupted by periods of tundra vegetation (Melisey I and II). The cold and dry periods are characterised by the development of a *Graminae–Artemisia* tree-less steppe vegetation in which several subarctic–subalpine species are evident. By contrast, the temperate St Germain episodes were associated with the development of a climax forest community.

The temperate conditions of St Germain II were abruptly terminated by a return to cold conditions associated with the Lanterne glaciation. Woillard (1978) subdivided this period of cold climate into three distinct cooling episodes. The first appears to have been interrupted by two short-lived periods of interstadial warmth. During the second episode, the cold and wet conditions were interrupted by two intervals of climatic warming. According to Woillard (1978), the first of these took place between 39,000 and 37,000 BP while the second occurred between 32,000 and 29,000 BP. The third episode of cooling corresponds to the last general glaciation when conditions were cold and dry.

129

Comparison of the Grand Pile record of vegetational change with the oxygen isotope stratigraphy obtained from ocean and ice cores (particularly the Camp Century Greenland curve of Dansgaard *et al.* 1971) indicates striking similarities (Mook and Woillard 1982) (Figure 2.6). For example, the Melisey cold intervals appear to correspond with isotope substages 5d and 5b while the transition between St Germain II and the first cycle of the Lanterne glaciation corresponds with the transition between oxygen isotope stages 5 and 4. Support for the validity of the Grand Pile succession has been derived from the pollen record at Les Echets near Lyon in central France (De Beaulieu and Reille 1984).

Les Echets: France

The Les Echets mire occurs 200km south of Grand Pile (De Beaulieu and Reille 1984). Here, the warm climatic conditions of the last interglacial were followed by temperate episodes during isotope substages 5c and 5a, each associated with the development of a deciduous forest cover. An older episode (substage 5c) is represented by two climatic optima separated by a brief climatic deterioration. However, periods of climatic cooling during oxygen isotope substages 5b and 5d are also indicated. Like Grand Pile, the time period represented by oxygen isotope stages 4, 3 and 2 (between approximately 70,000 and 10,000 BP) appears to have been characterised by cold and arid tundra conditions and interrupted on three (undated) occasions by periods of interstadial warmth and the development of a sparse forest cover. De Beaulieu and Reille (1984) noted that the vegetational response to climatic warming at the end of the last glaciation appears to have commenced near 15,000 BP when there was a marked rise in *Artemisia*.

Tenaghi-Phillippon: Greece

A long pollen curve of Late Quaternary vegetational changes similar to Grand Pile and Les Echets has also been obtained from northern Greece from a peat swamp at Tenaghi-Phillippon (Van der Hammen *et al.* 1971) (Figure 7.1). This record shows that arboreal taxa were characteristic of oxygen isotope stage 5 and were replaced by herbaceous taxa during isotope stages 4, 3 and 2. Notably, pronounced increases in herbaceous taxa and associated periods of cooling appear to correspond with isotope substages 5b and 5d. In addition, well-defined periods of climatic warmth occurred during substages 5e, 5c and 5a. The environmental deterioration that took place between stages 5 and 4 is particularly abrupt. Van der Hammen *et al.* (1971) showed that after the end of isotope stage 5, a marked increase in the proportion of tree pollen did not again occur until the transition between isotope stages 2 and 1. Van der Hammen *et al.* (1971) also noted that this climatic warming was interrupted and possibly reversed during the Younger Dryas, a factor not apparent in the records from Les Echets or Grand Pile.

Lake George: Australia

From Lake George, a long and continuous pollen record obtained from lake sediments is available for the environmental changes that took place since the last interglacial (Singh and Geissler 1985). The palaeobotanical data indicate that most of isotope stage 5 was dominated by an interglacial sclerophyll forest that was replaced during the isotope stage 5/4 transition by more herbaceous taxa. This time period is inferred to have been characterised by decreased temperatures, that subsequently increased during isotope stage 3. Similarly, the last glacial maximum was characterised by an absence of sclerophyllous taxa in the pollen spectra. Singh and Geissler (1985) suggested that low lake levels may have occurred in eastern Australia on only two occasions since the last interglacial – namely during the stage 5/4 transition and the culmination of isotope stage 2 – and that the intervening high lake levels were principally caused by reduced evaporation and increased precipitation.

THE SO-CALLED 'PLUVIAL' LAKES

Introduction

Until recently, it was widely believed that most lakes in non-glacial environments increased in area and volume during the last glacial period. The enlarged lakes were often referred to as 'pluvial' lakes although many factors in addition to precipitation may have affected their sizes. There has long been a controversy about whether wet 'pluvial' periods correspond with periods of glaciation in the middle and high latitudes. The factors that controlled the levels of 'pluvial' lakes are complex since they include rainfall, temperature, evaporation, relative humidity and groundwater conditions. As studies of 'pluvial' lake fluctuations have proceeded, it has become abundantly clear that, during the Late Quaternary, lakes in different parts of the world have responded to both changing regional and global climate conditions in a complex manner.

Pluvial lakes in tropical regions of Africa and America

East Africa

Detailed studies of lake sediment stratigraphy and geochemistry that have been undertaken for numerous East African lakes have shown that there has been a striking correspondence in the patterns of Late Quaternary lake level fluctuations for the region (Hamilton 1982) (Figure 7.2). The last glaciation, between 20,000 and 13,000 BP, was characterised by a prolonged phase of aridity that was preceded by a period of high lake levels that, in certain areas, may have lasted from at least 40,000 BP. In general, periods of low lake levels occurred before 12,000 BP as well as between 11,000 and 10,000 BP and between 8,000 and

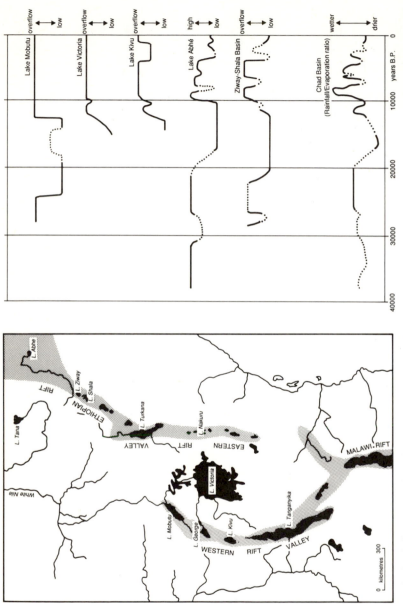

Figure 7.2 Lake level changes in selected basins in Africa. The basins are ordered from top to bottom approximately in order of increasing responsiveness to climate changes. Firm lines indicate those time periods where the lake level changes are firmly established. The locations of the individual lake basins are also shown (after Hamilton 1982).

7,000 BP (Street-Perrott and Perrott 1990). By contrast, lake levels were high between 12,000 and 11,000 BP and between 10,000 and 8,000 BP (Street-Perrott and Perrott 1990). The longest climatic record is for Lake Abhé (Gasse *et al.* 1980) in the Ethiopian Rift. Here, there is some evidence that the period of high lake levels that existed between 40,000 (perhaps as early as 50,000 BP) and 20,000 BP was interrupted by a brief phase of aridity at approximately 30,000 BP. Gasse (1977) and Gasse *et al.* (1980) have also provided some evidence that Lake Abhé may have dried out between 70,000 and 60,000 BP at a time when there may have been widespread glaciation in the northern hemisphere (see Chapter 4).

West Africa

The early lake level changes that are known for Lake Chad are in broad agreement with those derived for the East African lakes (Figure 7.2). In general, lake levels appear to have been high between 40,000 and 30,000 BP although interrupted by a brief dry phase near 30,000 BP. Thereafter, arid conditions prevailed between 20,000 and 12,000 BP when lake levels began to rise. During the maximum phases of lake expansion, Lake Chad was approximately 1,100 km in length and 680 km wide (Grove and Warren 1968). The rise in lake level was, however, interrupted during the Younger Dryas (Servant and Servant-Valdary 1980).

Tropical America

During the last glacial maximum, lake levels were low in equatorial America (Flenley 1979). However, a series of rapid lake level fluctuations took place soon after 14,000 BP (Street-Perrott and Perrott 1990). Data from several lakes in Mexico and Colombia indicate that lake levels rose rapidly after approximately 12,900 BP (Flenley 1979; Street-Perrott and Perrott 1990). Thereafter lake levels fell before rising again between 8,000 and 7,000 BP. Street-Perrott and Perrott (1990) noted that the Mexican pattern of lake level changes was broadly similar to those in Africa but differed from them in detail, particularly in that there was no evidence for a high lake stand between 10,000 and 8,000 BP.

Cause: changes in the production of North Atlantic deep water

Street-Perrott and Perrott (1990) have provided a very interesting hypothesis to explain the observed sequence of tropical lake level fluctuations. They suggest that periods of African aridity correspond with periods of anomalous sea surface temperatures in the North Atlantic. They attributed these periods to times when large volumes of glacial meltwater entered the North Atlantic following the emptying of large ice-dammed lakes developed along the southern margins of the Laurentide ice sheet. During such periods (e.g. between 11,000 and 10,000 BP and soon after 8,000 BP) the increase in ocean salinity stratification that was associated with reduced North Atlantic deep water (NADW) formation and lowered sea surface temperatures led indirectly to decreased rainfall and low lake levels in the tropical areas of Africa and America. Conversely, during periods when the flow of Laurentide meltwater took place into the Gulf of Mexico (e.g. between 13,500 and 12,000 BP and between 9,600 and 8,000 BP) the production of NADW was resumed and lake levels began to rise.

It is not immediately clear why lowered sea surface temperatures in the North Atlantic should have led to aridity in tropical Africa and perhaps elsewhere in the low latitudes. One view is that it was caused by a reduction in the northward oceanic transport of heat across the Equator. This would have taken place when there was a marked north (cold) – south (warm) contrast in sea surface temperatures (Street-Perrott and Perrott 1990). Thus, during the last glacial maximum, when the North Atlantic had a cover of sea ice, the production of NADW was halted and arid conditions prevailed in tropical Africa and America.

According to this hypothesis, the influence of glacial meltwater on NADW formation would have varied greatly according to the location at which the fresh water was discharged into the ocean. At present in the North Atlantic, oceanic upwelling during winter leads to widespread chilling of the surface waters. The subsequent sinking of this cold saline water leads to the formation of NADW. During a glacial age, the influx of fresh water to the North Atlantic reduced the salinity of the surface waters. The resulting decrease in the density of the surface waters therefore prevented them from sinking and hence halted the formation of NADW. This interpretation has been given additional support by the observation that recent periods of drought in sub-Saharan Africa appear to correlate with periods when there have been strong sea surface temperature anomalies in the central Atlantic (Street-Perrott and Perrott 1990).

'Pluvial' Lakes in Mid-Latitude USA

During the last glaciation in the western United States numerous lakes were produced in Utah, Nevada and Oregon as well as several in eastern California, New Mexico and Texas. Of these, the largest were Lake Bonneville in Utah and Lake Lahontan in western Nevada (Figure 7.3) (Plate 7). Lake Bonneville was the largest Late Quaternary pluvial lake in North America with an area of 52,000 km^2 and a volume of 7,500 km^3 (Gilbert 1890; Crittenden 1963; Smith and Street-Perrott (1983). Recent research (Currey 1980) appears to indicate that progressive filling of the lake took place from 25,000 BP until maximum volumes were reached between 18,000 and 15,000 BP. Smith and Street-Perrot (1983) have suggested that during the last glacial maximum, fluvial maximum volumes were reached between 18,000 and 15,000 BP. Smith and Street-Perrot (1983) have suggested that during the last glacial maximum, fluvial discharge into the lake was at least five times higher than at present and that evaporation was probably between 10 per cent and 40 per cent lower than

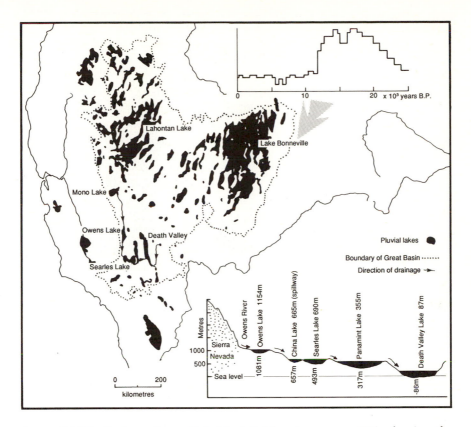

Figure 7.3 Distribution of so-called 'pluvial' lakes in western USA showing the chronology of Lake Bonneville lake level fluctuations for the last 20,000 years. The cascading flow from Mono Lake that affected the levels of several other 'pluvial' lakes in SW USA is also indicated (after Smith and Street-Perrott 1983).

present (Figure 7.3). Between 15,000 and 10,000 BP, however, there was a progressive fall in lake levels. After the latter date, the lake appears to have been completely desiccated.

During the last glaciation, Lake Lahontan was produced in northwest Nevada as a complex of interconnected water bodies approximately half the size of Lake Bonneville (Russell 1885). Numerous studies have been undertaken in order to reconstruct the environmental history of the lake (e.g. Broecker and Orr 1958; Benson 1981) and it appears from these that the highest lake levels occurred at 15,000 BP and that, thereafter, desiccation and progressive lake lowering took place until 10,000 BP. The early history of Lake Lahontan is uncertain although lake levels appear to have been increasing from 22,000 BP until 15,000 BP (Smith and Street-Perrott 1983) (Plate 8).

Farther south, in southeastern California, glacial meltwater from the Sierra

Plate 7 Shorelines of Lake Bonneville. Photo K. Richards.

Nevada Range discharged westwards and cascaded through a series of pluvial lakes, the highest of which was Owens Lake and with the lowest occupying Death Valley (Figure 7.3). Overflow from Mono Lake to the north also contributed to discharge into Owens Lake from the Sierra Nevada glaciers. Mono Lake reached its highest level at 13,000 BP and thereafter was characterised by lowering until levels similar to the present were reached by 10,000 BP. Overflow from the lake drained through the Owens River valley into Owens Lake which in turn received additional discharge from the Sierra Nevada. The overflow from Owens Lake, in turn, cascaded through Searles Lake, Panamint Lake and ultimately into Death Valley Lake (Smith and Street-Perrott 1983) the altitude of which (+87m) contrasts with that of the present desiccated surface (−86m) (Hooke 1972) (Figure 7.3). Smith and Street-Perrott (1983) concluded that between 25,000 and 10,000 BP, this region was characterised by marked reductions in both air temperatures (5–10°C) and evaporation.

Smith and Street-Perrott (1983) concluded that during the Late Quaternary in the western United States the highest lake levels (with the exception of Lake Bonneville) mostly occurred between 24,000 and 21,000 BP while lake levels declined, although they still remained high, between 21,000 and 14,000 BP. Thereafter, lake levels appear to have declined with the exception of a distinct period of lake expansion during the Younger Dryas (Smith and Street-Perrott 1983). By contrast in the eastern United States, the Late Wisconsin glaciation appears to have been characterised by increased aridity and hence there is no

Plate 8 Lacustrine sediments of Lake Lahontan at Rye Patch. Photo C. Embleton.

evidence for the former existence of large pluvial lakes in this region although decreased evaporation and lowered temperatures are thought to have resulted in the development of several small lakes (Barry 1983).

Cause: jet stream displacement

It is generally agreed that the high 'pluvial' lake levels that existed in south-western USA during the last glaciation resulted from the combined influence of increased precipitation, decreased evaporation, lower temperatures, as well as the local supply of meltwater from alpine valley glacier complexes. The meteorological changes were principally the result of increased cyclogenesis over the south-western United States. This appears to have been due to the southward steering of mid latitude cyclones around the southern margin of the Laurentide and Cordilleran ice sheets. These processes were, however, complicated by patterns of Late Wisconsin adiabatic descent of air in the lee of the Pacific mountain chains.

In eastern USA, however, the inferred reductions in both precipitation and temperature appear to have resulted from air subsidence, and hence enhanced anticyclonic conditions. These processes may have been due to a westerly jet stream that was displaced southwards of the Laurentide ice sheet and whose cyclonic curvature may have induced high pressure near the ground surface (Barry 1983; Kutzbach and Guetter 1986).

Pluvial lakes in Australia

Bowler (1978) has summarised the evidence for Late Quaternary lake level fluctuations in Australia where, as in Africa, lake levels appear to have been high prior to the last glacial maximum but started to fall after 25,000 BP. The last glacial maximum appears to have been characterised by aridity and low lake levels, with the period of maximum aridity having occurred between 18,000 and 16,000 BP. Thereafter, lake levels began to rise at 13,000 BP and had reached high levels by 9,000 BP before progressively falling throughout the remainder of the Holocene.

Cause: strengthened anticyclones

The extreme aridity of the last glacial maximum in Australia may be explained principally by enhanced anticyclonic conditions over the continental interior that would have become more intense due to lower ground surface temperatures (Gates 1976a,b). In addition, lower sea surface temperatures are likely to have moderated the development of tropical cyclones over northern Australia (Wyr-woll and Milton 1976). The distribution of sea ice in the Southern Ocean during the last glacial maximum is also crucial in this regard since, through its influence on Antarctic high pressure, it was instrumental in defining the tracks of mid latitude cyclones over the Southern Ocean. These processes in conjunction with glacio-eustatic sea level lowering and increased continentality may have resulted in greater seasonal contrasts in Australia during the last glacial maximum in which cool stormy winters alternated with hot dry dusty summers (Bowler 1978).

The Middle East

Farrand (1971) has described a series of pluvial lakes in the Middle East that formed during the last glacial period. The largest of these were located within the Dead Sea Rift Valley. During this period, a large lake (Lake Lisan) formed in this area and is estimated to have had a volume of 325 km^3. The main period of lake expansion is thought to have occurred between 18,000 and 12,000 BP (Rognon 1976). However, reconstructions of former lake volume and of former precipitation and evaporation values is hindered by a complex history of neotectonic activity.

Cause: jet stream displacement

The principal cause of high lake levels in the Middle East during the last glaciation was a southward displacement of the mid latitude jet stream due to a permanent anticyclone that formed over the Eurasian ice sheet and central Europe (Figure 3.2). The numerical models of Kutzbach and Wright (1985) show that the mid latitude jet stream may have split around the Eurasian ice sheet with the southern limb having tracked eastwards across the Mediterranean. Under such circumstances, reduced temperatures, increased rainfall and decreased evaporation may have led to the widespread development of lakes (Nicholson and Flohn 1980) (Figure 7.4). Furthermore, such conditions are likely to have prevailed until the final disappearance of the Eurasian ice masses during the early Holocene.

ICE-DAMMED LAKES

It has always been accepted that several extremely large ice-dammed lakes were impounded against the margins of the last ice sheets in the northern hemisphere. At present, those former ice-dammed lakes known to have existed are those where the stratigraphic and geomorphological evidence presents compelling evidence for their former occurrence. Not surprisingly, the controversy about the former existence or otherwise of the large proglacial Pur and Mensi ice-dammed lakes in the Soviet Union during the last glaciation (see Chapter 5) is centred upon the question of whether or not the geological evidence is compatible with such a view (Grosswald 1980). Elsewhere, large lakes are known to have been impounded along the southern margins of the Laurentide, Cordilleran and Scandinavian ice sheets (e.g. glacial Lake Agassiz, glacial Lake Missoula and the Baltic Ice Lake) (see Chapter 5).

During the melting of the last northern hemisphere ice sheets, numerous catas-

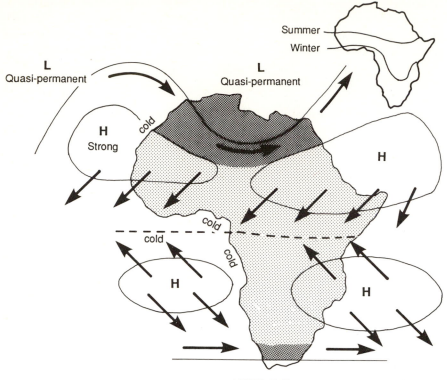

18000 B.P.

Figure 7.4 Inferred pattern of atmospheric circulation over Africa during the last glacial maximum. Dark shading indicates those areas more humid than today; lighter shading indicates areas considered drier than today. The inset shows the present position of the ITCZ in summer and winter months (after Nicholson and Flohn 1980; Bradley 1985).

trophic floods took place in conjunction with the drainage of these (and other) lakes. Indeed, the drainage of the largest lakes may have been sufficiently large to have resulted in almost instantaneous rises in global sea level (see Chapter 8). Under such conditions where mass is rapidly redistributed over the Earth's surface, earthquakes and tectonic activity are very likely to have taken place. Although little is known about the evolution of the largest lakes, sufficient information is available to indicate that their sudden emptying may have been associated with widespread seismicity and crustal deformation (see Chapter 11).

SUMMARY

Studies of lakes, bogs and mires that existed during the Late Quaternary have provided valuable information regarding past environmental changes. Investigation of long sediment sequences has shown that the pattern of former environmental changes exhibits great regional complexity and variability. One point that emerges clearly is that there was a pronounced deterioration in global climate during the transition between oxygen isotope stages 5 and 4. This change is evident from sites as far apart as those in Australia, France and Colombia. The climatic deterioration was associated not only with a decrease in temperature but also by marked vegetational changes in which there was a widespread decrease in the deposition of tree pollen and an increase in the proportions of herbaceous taxa.

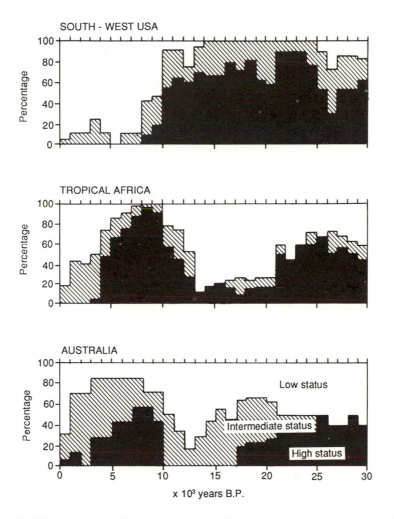

Figure 7.5 Histograms showing lake level status for thousand year time intervals from 30,000 BP to the present day for three areas: southwestern USA, intertropical Africa and Australia (after Street and Grove 1979).

It is difficult to ascertain the climatic changes that took place between the isotope stage 5/4 transition and stage 2. The model of North Atlantic deep water (NADW) formation described by Street-Perrott and Perrott (1990) might provide some important clues since it establishes a link between phases of tropical African aridity, rates of glacial meltwater discharge into the North Atlantic and patterns of oceanic convection. If this is correct, the data provided by Gasse *et al.* (1980) for lake level changes in Ethiopia lend support to the view that the periods of African aridity centred on 70,000 BP and 30,000 BP may broadly equate with periods of glaciation in North America.

The lake level fluctuations that took place during the last 30,000 years highlight striking regional variations and demonstrate quite clearly that there were no 'pluvial' lakes in the traditional sense. For example, the patterns of lake level fluctuations exhibit complex variations not only between North America, Africa and Australia but also within individual continental areas (Figure 7.5). Thus, in the western USA, lake levels were high during the last glacial maximum, while they were low in eastern North America. The reason for such a marked contrast was most likely due to the southward displacement of the mid latitude jet stream due to the presence of the Laurentide ice sheet. In western USA, the pattern of jet stream flow may have led to the frequent development of cyclonic depressions, which coupled with lower temperatures and rates of evaporation led to high lake levels. By contrast, in the eastern USA, jet stream flow may have been associated with the more frequent occurrence of anticyclonic conditions. Increased cyclone activity coupled with jet stream displacement may also be invoked to account for the relatively high lake levels that occurred in North Africa during the last glaciation.

Several large ice-dammed lakes were formed at the margins of the Laurentide, Cordilleran and Eurasian ice sheets during the last glaciation. All of these lakes were relatively short-lived phenomena and were eventually emptied due to ice thinning and retreat. According to Street-Perrott and Perrott (1990) the periodic drainage of the largest of these lakes, glacial Lake Agassiz, may have induced global oceanographic and atmospheric changes that led to major lake level fluctuations elsewhere in the world. It remains to be discovered if the sudden redistribution of mass associated with some of these lake drainage events led to extensive earthquake and tectonic activity.

8

RIVERS

INTRODUCTION

In 1909, Penck and Bruckner published the results of their research on alpine glaciations. They had observed that a series of river terraces is present in the principal valleys draining the Alps. They made a clear distinction between the periods of fluvial aggradation and incision that led to the development of the series of terraces. Stratigraphic studies of the terrace gravels, in conjunction with studies of fossil soils buried within the terraces, led them to conclude that the periods of fluvial aggradation coincided with periods of glacier expansion and meltwater deposition. The periods of soil development and river incision were attributed to warm interglacial episodes. This research, based largely on the interpretation of fluviatile sediments, led to the establishment of the fourfold scheme of Quaternary glaciation. The phases of glaciation were named, in descending age order, Gunz, Mindel, Riss and Wurm after the major rivers in southern Germany on which their observations were made.

Later research showed that each cold episode could be divided into several substages and these, in turn, implied the former occurrence of a very complex series of past climatic fluctuations. Indeed, it was shown that some of the phases of fluvial incision may have taken place during periods of cold, rather than warm, climate. The immense importance attributed to the Penck and Bruckner model of glaciation throughout much of the 20th century highlights the dangers in assuming that rivers respond in a straightforward way to long-term climatic changes. It is now well known that episodes of high sediment yield that cause fluvial aggradation may be due to such diverse causes as glacial meltwater flow, rising base level, tectonic uplift in the source area, sea level changes, as well as climatic change itself (Baker 1983a,b). The complex relationship between river terrace development and secular environmental changes is well illustrated by the scheme of terrace development of the River Thames described by Green and MacGregor (1980) (Figure 8.1) in which the form of individual terraces disguises numerous different stratigraphic sequences.

The hydrological response of rivers to climatic changes during the Late Quaternary may thus have been extremely complex due to the diffusion of

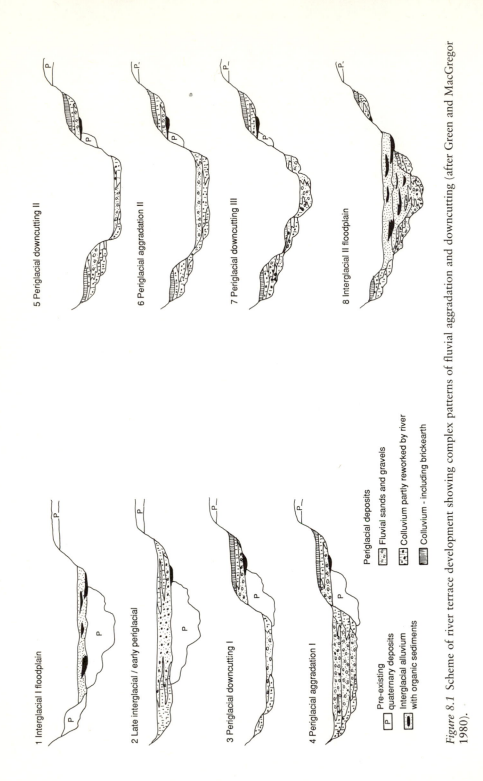

Figure 8.1 Scheme of river terrace development showing complex patterns of fluvial aggradation and downcutting (after Green and MacGregor 1980).

imposed changes through individual river systems. Furthermore, the complexity of response may have been compounded by the existence of critical extrinsic and intrinsic thresholds (Schumm 1974). For example, extrinsic thresholds may have been triggered by climatic changes that led to vegetational changes that, in turn, induced dramatic changes in the patterns of fluvial erosion and deposition. By contrast, intrinsic thresholds may have been exceeded by changes within individual floodplains (e.g. changes in slope that led to periods of accelerated erosion or multiple phases of fluvial aggradation due to a single external stimulus (Schumm 1974; Patton and Schumm 1981)).

There are two principal ways in which it is possible to establish the response of rivers to Late Quaternary environmental change. The most common approach has sought to establish former run-off characteristics on the basis of detailed sedimentological and stratigraphical investigations. The information on inferred patterns of former discharge has then been used to reconstruct past patterns of discharge that have, in turn, been used to build models of regional climatic change. In many cases, models of modern river dynamics have been used to establish criteria from which estimates can be made of the likely critical response

Table 8.1 Summary of changes in channel behaviour (erosion–deposition) during climate change

Location or type of basin	Climate change to	Dominant effect	River activity
1. Basin partly occupied by ice sheet	Cooler, wetter or cooler, drier	Increased sediment load	Deposition
2. Basin partly occupied by valley glacier	Cooler, wetter or cooler, drier	Increased sediment load	Deposition
3. Periglacial	Cooler, wetter or cooler, drier	Increased sediment load	Deposition
4. Coastal	Cooler	Fall of sea level	Erosion
5. Interior basin	Cooler, wetter	Rise of lake levels	Deposition
6. Unglaciated continental interior humid, perennial streams	Cooler, wetter	Increased water discharge Minor increase of sediment yield	Erosion
7. Unglaciated continental interior semi-arid, ephemeral streams	Cooler, wetter	Significant decrease of sediment yield	Erosion
8. Unglaciated continental interior arid, ephemeral streams (a) tributaries (b) main rivers	Cooler, wetter	Significant increase of water discharge Increase of drainage density Significant increase of sediment load	Erosion Deposition

After Chorley et al. *(1984)*

of rivers to particular environmental changes (Table 8.1). A different approach has attempted to develop a general model of river basin response that is extrinsically driven by climatic forcing functions (Baker 1983a,b; Thornes 1987). In this chapter, the palaeohydrological response to former environmental changes is illustrated from both points of view. Particular attention is given to extreme fluvial 'events' since these not only represent individual hydrological responses to global environmental change but also, in some cases, may have been of sufficient magnitude in themselves to have modified global climate.

FACTORS THAT AFFECT HYDROLOGICAL RESPONSE

Schumm (1965) and Chorley *et al.* (1984) proposed that changes in Quaternary river activity took place as a result of a series of different climatic changes (Table 8.1). They argued that the hydrologic response of a river throughout a glacial–interglacial cycle would vary according to the location or type of the drainage basin. Thus, a drainage basin located in the continental interior would be hot and dry during interglacials and cool and wet during glacials. Accordingly, the river response would fluctuate between deposition at the end of interglacials and fluvial erosion at the end of glacial periods and in the early phases of interglacials (Table 8.1).

A major factor affecting the response of mid latitude rivers to the ending of glaciation was the melting of permafrost. This led, through the invasion of forest communities and the establishment of a vegetation cover, to marked decreases in river discharge and sediment load. In many areas there appears to have been an initial change from braided to meandering river systems and a later change from large to small palaeomeanders (Briggs and Gilbertson 1980; Starkel 1987, 1990). In many areas the changeover from braided to meandering regimes coincided with the end of high discharge events caused by the final melting of ice sheets. In addition, marked hydrological changes appear to have taken place throughout the period of overall ice sheet decay prior to final melting. Thus Starkel (1983) has argued that, whereas the Alleröd interstadial in Central Europe was dominated by incision and meandering rivers, the later Younger Dryas period of cold climate was accompanied by increased braiding and the deposition of coarser material.

In many areas affected by tectonic uplift during the Late Quaternary, river gradients have become steeper and have induced active downcutting that has led to the formation of river terraces. Similar complex responses also appear to have taken place in areas where glacio-isostatic uplift has taken place due to the melting of ice sheets. In the lower courses of many rivers, the rise in world sea level of circa 120m that has taken place since the last glacial maximum has led to the shortening of many rivers and an increase in valley aggradation.

The most important factor, however, affecting hydrological response has been climate. For example, changes in rainfall and potential evapotranspiration affect soil moisture and vegetation growth processes. These, in turn, are important

146

influences on evapotranspiration, as well as on surface run-off and groundwater discharge. As a result, many rivers were subject to 'metamorphoses' (Schumm 1977) as the palaeochannels experienced large changes in discharge and sediment load, as well as occasional diversion of drainage and/or episodes of cataclysmic flooding (Baker 1983a,b). Baker (1983a,b) has drawn attention to the occurrence of abnormally high discharge events or 'superfloods' that have taken place during the Late Quaternary and has noted that the results of these have been frequently preserved in the present landscape.

Most of the inferences that have been made about the nature of hydrological changes during and after the last ice age are based on local sedimentological, stratigraphical and geomorphological investigations. These have been used to reconstruct the local and regional-scale controls that have affected river response. Such studies have highlighted the often dramatic changes in river regime that took place, most of which are only known at present in the barest outline.

PALAEOHYDROLOGICAL CHANGES IN LOW LATITUDES

In Africa, much attention has been given to the palaeohydrology of the Nile (Adamson and Williams 1980; Butzer 1980; Williams and Adamson 1980) (Figure 8.2). It is widely believed that the Nile drainage basin was extremely arid throughout much of the time period from 25,000 BP until about 12,500 BP. In Ethiopia, the catchment area of the Blue Nile appears to have been characterised by a sparse vegetation cover during the last glacial maximum. Rognon (1980) also concluded that during this period the Blue Nile was characterised by seasonal braided river flow due to a reduced vegetation cover.

Throughout most of this period there appears to have been very reduced river discharge into the Nile from the lakes in East Africa (e.g. Lake Victoria and Lake Albert (Lake Mobutu Sese Seko)). Livingstone (1980) has argued that Lake Albert, although open to the Nile between 28,000 and 25,000 BP, did not overflow into the Nile headwaters between 25,000 and 18,000 BP. Although overflow later took place between 18,000 and 14,000 BP, the principal flooding took place soon after 12,500 BP. In the case of Lake Victoria, overflow of waters into the Nile also did not commence until this time not only as a result of rising lake levels but also due to tectonic uplift along parts of the Nile watershed (Figure 8.2). Butzer (1980) has described a similar pattern of environmental changes for the Nile Valley in Egypt and Lower Nubia. He concluded that the climate was hyperarid between 25,000 and 18,000 BP although the presence of flood silts from the Ethiopian Highlands is indicative of some seasonal flooding at this time. Between 18,000 and 17,500 BP severe aridity resulted in considerable fluvial dissection and locally led to the invasion of floodplains by dune fields. Thereafter, several episodes of increased flooding took place between 14,500 and 12,500 BP with severe flooding between 12,000 and 11,500 BP. During these periods, the hydrological response of the Nile was complex with the transport of

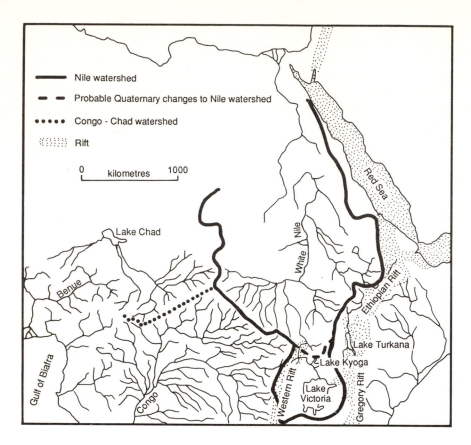

Figure 8.2 Drainage patterns in northeastern and central Africa showing the present and probable former limits to the Nile basin (after Adamson and Williams 1980).

large quantities of summer flood silts from Ethiopia as well as substantial accumulations of bed load sediments.

The great flood events that took place along the Nile between 12,000 and 11,500 BP have been remarked upon by several authors (e.g. Butzer 1980; Livingstone 1980; Hamilton 1982). In Lower Nubia the floods may have reached as much as 9 m above the floodplain of the time and are thought by Butzer (1980) to have reflected unusual climatic anomalies in sub-Saharan Africa at this time. There are two reasons why increased flooding may have taken place during this period. One possibility is that the SW monsoon in eastern Africa began to function again after it had previously failed during the last glacial period. The resumption of monsoonal circulation, and the related changes in the position of the Intertropical Convergence Zone, may have been associated with the reorganisation of global atmospheric circulation that accompanied the rapid thinning and retreat of ice sheets in the northern hemisphere.

An alternative hypothesis is that the fluctuations in aridity resulted from changes in precipitation and wind vigour and that these changes took place independently of any changes in the position of the Intertropical Convergence Zone. The different climatic models that can be invoked to account for such changes highlight the way in which an inductive approach to palaeohydrology can lead to a variety of different palaeoclimatic scenarios.

In the Amazon Basin, the last glacial maximum was characterised by a substantial reduction in the area of tropical rainforest due to changes in the position of the Inter Tropical Convergence Zone. Van der Hammen (1974) has provided palynological data demonstrating that the Basin was affected by drier and colder climates while Damuth and Fairbridge (1970) have provided sedimentary evidence from the continental shelf for increased aridity in the Amazon Basin during the last glacial maximum. However, since much of the present rainfall in the basin is provided by rainforest evapotranspiration (Friedman 1977) it is not surprising that fluvial discharge in the Amazon Basin during the last glacial period was considerably lower than present, because of the disappearance of large areas of rainforest (see Chapter 3).

Similar glacial-age aridity may have occurred in western Africa where investigations of offshore ocean sediment cores has shown that pronounced aridity occurred between 19,000 BP and 12,500 BP (Rossignol-Strick and Duzer 1980). Under such circumstances, fluvial activity in the West African drainage basins may have been drastically reduced during the last glacial maximum (Sarnthein and Koopman 1980). However, Talbot (1980) has shown that fully integrated drainage networks may have later become active in West Africa between 12,000 and 8,000 BP and may have extended as far north as the Tropic of Cancer (Figure 8.3). Talbot considered that most sediment deposition, in association with a reduced vegetation cover, took place along braided rivers and that most sediment accumulated when rainfall was generally quite high yet rather seasonal.

The cause of the changes in precipitation that led to the development of river networks in West Africa is not clear. It seems likely that the increased aridity during the last glacial maximum may have resulted from a failure of the SW monsoon. Equally, it is possible that the more seasonal and more extensive rainfall between 12,000 and 8,000 BP might have been due to a contracted Intertropical Convergence Zone that was displaced slightly farther south during winter and farther north during the summer monsoon. However, the patterns of monsoonal rainfall are also likely to have been influenced by changes in temperature, relative humidity and vegetation cover. The uncertainties again illustrate the severe problems that one encounters when attempting to reconstruct patterns of climate change from local sedimentological and stratigraphic evidence.

Limited information is available from China on the response of the Yangtse and Huang Ho rivers to climatic changes during the Late Quaternary. Huairen et al. (1986) and Huairen and Xi-qing (1987) have shown that as a result of the low sea levels (between −120 and −130 m) in eastern China during the last glacial

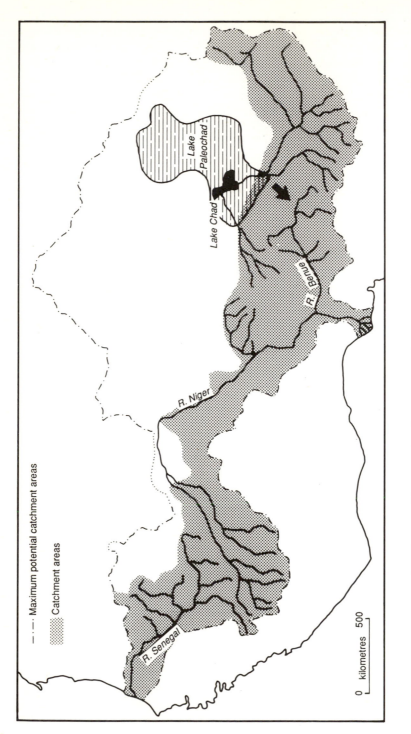

Figure 8.3 Maximum potential catchments of the Senegal and Niger–Benue river systems. Those areas of catchment presently contributing run-off are shaded. During periods of maximum humidity, particularly during the early Holocene, all of the catchments were probably active. Overflow from the Chad Basin reached the Benue via the arrowed route (after Talbot 1980).

Legend:
— · — · — Maximum potential catchment areas

Catchment areas

Lake Paleochad

Lake Chad

R. Benue

R. Niger

R. Senegal

0 kilometres 500

maximum, the Yangtse was subject to headward erosion and valley incision that reached as far as 1000 km from the present estuary. However, the response of the major Chinese rivers has also been greatly influenced by Late Quaternary neotectonics that have largely controlled patterns of fluvial aggradation and downcutting.

SUPERFLOOD 'EVENTS' AND ICE-DAMMED LAKES

Processes

Lakes that are impounded by glacier ice may remain relatively stable for long periods of time. However, when the ice dam is weakened, usually by ice thinning and retreat, the lake may drain suddenly (Clague and Mathews 1973). Frequently, catastrophic outburst floods (jökulhlaupe) will commence in this way when the forces exerted by the water exceed the strength of the dam. When this happens, the ice dam may become buoyant, enabling the discharge of lake water beneath the ice along subglacial tunnels. These subglacial floodwaters rapidly enlarge the tunnels until most of the lake is drained. Thereafter the tunnel will close itself through confining ice pressures and will eventually lead to the refilling of the lake. In some cases, the outburst flood may be so great that the ice barrier completely disintegrates, thus preventing the lake from ever forming again. Ice-dammed lakes that are impounded by cold-based ice sheets are more stable features since it takes much longer for the ice dam to become buoyant. Under such circumstances, the overflow waters from the lake are continually discharged from the lake across the lowest topographic area surrounding the lake.

Studies that have been undertaken on recent catastrophic floods caused by the breaching of ice barriers have shown that they exhibit a distinctive pattern of discharge. In general there is an exponential increase in flow during the rising stage followed by a brief discharge maximum. Thereafter there is a rapid decrease in flow as the supply of water from the draining lake decreases (Clague and Mathews 1973). It is believed that catastrophic outburst floods from ice-dammed lakes during the last ice age were characterised by discharge patterns analogous to those of the present although of considerably greater magnitude.

Superfloods

Many of the rivers that drained the major ice sheets experienced extremely high discharge (superfloods) (Baker 1983a,b). Broadly, there are two different types of superflood. The first consists of periods of greatly increased discharge resulting from the drainage of glacial meltwaters from ice sheets (mostly during the ablation season). The second arises due to the formation of large ice-dammed lakes adjacent to ice sheets. Under such circumstances flooding can be caused by the overflow of lake waters into the headwaters of a drainage basin. In addition, rapid flooding of glacial meltwaters into ice-dammed lakes often leads to increased rates of overflow. Finally, catastrophic emptying (jökulhlaup) of glacier or ice-sheet-dammed lakes can cause very severe flooding.

Laurentide and Cordilleran superflood 'events'

Superfloods related to the Cordilleran ice sheet

Catastrophic floods are inferred to have taken place due to the periodic drainage of glacial Lake Missoula in Washington and Idaho (see Chapter 5) which, when dammed, contained $2,000 \, km^3$ of water at its maximum (Baker 1983a,b; Waitt 1985). These catastrophic floods are considered to have eroded the basalt and loess areas of the Channeled Scablands of the Columbia River Basin, Washington, as well as having deposited considerable volumes of sediment on the continental shelf and slope (Bretz 1969). The approximate timing of the last floods is shown by the presence of Mt St Helens tephra of 13,150 BP within Missoula flood sediments (Baker and Bunker 1985). Baker (1983a,b) has estimated that the peak discharge associated with the catastrophic drainage of glacial Lake Missoula may have been approximately $21 \times 10^6 \, m^3 \, s^{-1}$ – an astonishing value that is approximately 20 times greater than the average present worldwide run-off of 1.1 million $m^3 \, s^{-1}$. Waitt (1980, 1985) has suggested that as many as 40 catastrophic floods may have affected the Channeled Scabland area since the lake may have taken about 100 years to refill after each lake-emptying event (Plate 4).

The Missoula floods led to the development of a number of distinctive fluvial landforms. For example, extremely large boulder bars were deposited in many areas while individual hills of loess were streamlined by flood flows. Elsewhere, basalt bedrock was eroded to produce complex systems of erosional grooves, potholes, rock basins and cataracts while giant current ripples, locally over 5 m high and spaced 100 m apart were also deposited (Baker 1983a,b; Waitt 1985).

Superfloods related to the Laurentide ice sheet

Mississippi superfloods

The different types of superflooding are well illustrated by the patterns of fluvial discharge through the Mississippi–Missouri–Ohio drainage basin. During the Late Wisconsin, this basin received meltwater from a 2,700 km sector of the Laurentide ice sheet margin (Baker 1983a,b). The pattern of drainage during the last 25,000 years can broadly be divided into four separate episodes (Table 8.2).

(i) The first phase corresponds with the period prior to the last glacial maximum when the development of braided streams was widespread. The pattern of discharge was probably influenced by the fact that large areas of the Laurentide ice sheet were cold based. Because of this, the principal supply of meltwater was seasonal due to summer ice ablation and snowmelt. The lower

Table 8.2 Late Quaternary fluvial events in the Mississippi River drainage system

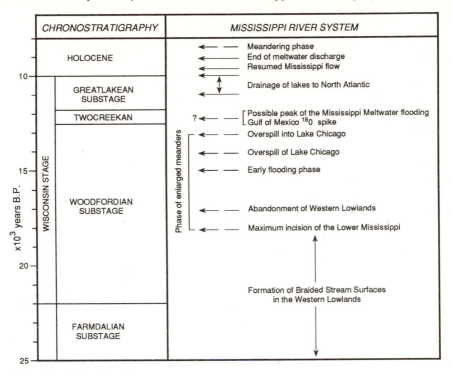

Adapted from Baker (1983b)

reaches of the Mississippi valley were considerably influenced by glacio-eustatic sea level lowering that may have reached a maximum value of $-120\,\mathrm{m}$ at 18,000 BP. This resulted in widespread valley incision and the development of a Mississippi delta in the Gulf of Mexico many hundreds of kilometres south of its present position.

(ii) The second episode was characterised by great increases in river discharge and the aggradation of very extensive outwash plains in the upper Mississippi Basin (Baker 1983a,b). Baker suggested that this period of rapid aggradation coincided with the early stages of ice sheet deglaciation due to the combined influences of high sediment loads and a rising sea level.

(iii) The third period corresponds to the time interval when, due to ice retreat, large ice-dammed lakes formed in the headwaters of the drainage basin and provided discharge through lake overflow. On numerous occasions increased flooding of waters into these lakes led to corresponding increases in the rate of discharge of meltwater through the drainage basin. These events may have been accompanied by a decrease in sediment aggradation since the presence of the lakes permitted only finer-grained sediments to be transported downstream (Baker 1983a,b). However, the greater rates of river discharge led to increased erosion and transport of glacigenic and fluviatile sediments already contained within the catchment.

The first major ice-dammed lake (Lake Chicago) formed at 14,000 BP while the last (glacial Lake Agassiz–Ojibway) was not emptied until 8,000 BP. The key factor, however, that influenced the palaeohydrological evolution of the Mississippi drainage basin during this period was the development of glacial Lake Agassiz that was subjected to a series of fluctuations in size between 12,800 BP and 8,000 BP.

The hydrological response to glacial Lake Agassiz

Throughout the existence of glacial Lake Agassiz remarkable fluvial discharge occurred through the Minnesota spillway into the Mississippi drainage basin. For example, large volumes of meltwater were supplied through River Warren, the southern outlet of glacial Lake Agassiz, as it merged with the Mississippi. Matsch (1983) has estimated that the bankfull discharge related to Lake Agassiz overflow waters in River Warren during periods of stable geometry was 40,000 m^3 s^{-1}, while higher rates of flow accompanied periods of lake flooding when large volumes of meltwater drained from the ice sheet into the lake. Matsch (1983) concluded that these periodic floods may have been associated with peak discharges of 1,000,000 m^3 s^{-1}. He also concluded that 75,000 km^3 of drift and bedrock may have been removed from the course of this river across Minnesota and that it may have taken nearly 500 years to flush the Minnesota portion of the River Warren sediments into the Gulf of Mexico as part of the Mississippi River suspended sediment load.

Indirect evidence for a period of very high river discharges into the Gulf of Mexico through the Mississippi is provided from the oxygen isotope stratigraphy of the Gulf of Mexico sea floor sediments where enhanced ^{18}O depletion (i.e. increased freshwater influx) is evident between approximately 12,500 and 11,500 BP (Broecker et al. 1989) (Figure 5.8 inset). Relatively high Mississippi discharge also occurred soon after 10,000 BP and may have continued until near 9,000 BP as meltwater from Lake Agassiz continued to overflow into the Mississippi Basin. Reduced flow is thought to have occurred, however, between 11,000 and 10,000 BP when ice retreat may have diverted meltwaters from Agassiz eastwards to Lake Superior and along the St Lawrence into the North Atlantic (Figure 5.8) (Plate 5).

(iv) The supply of water from Lake Agassiz to the Mississippi River Basin was ended about 9,200 BP. Soon afterwards, there was a change from a braided to a meandering regime in the Mississippi valley with alternating brief episodes of fluvial aggradation and dissection (Baker 1983a,b). The processes of fluvial aggradation within the lower Mississippi valley were particularly favoured due to the continued rise in sea level that did not reach its present level until near 6,500 BP.

Other Laurentide superfloods

On several occasions during regional deglaciation, waters from Lake Agassiz breached confining ice barriers and drained as a series of catastrophic floods. For example, major jökulhlaupe occurred at 11,000 BP due to the escape eastwards of meltwaters through Lake Superior and the St Lawrence Lowlands into the North Atlantic. A second series of major flood 'events' took place at 9,400 BP when escaping meltwaters followed a broadly similar route eastwards into the North Atlantic (Teller 1990; Teller and Thorleifson 1983) with typical discharge values of 100,000 m^3 s^{-1} (Figure 8.4). Teller and Thorleifson (1983) have estimated that about 4,000 km^3 of water were discharged during each flood event, a volume more than nine times that of the present Lake Erie, and twice the flood volume from the Lake Missoula jökulhlaupe that resulted in the development of the Channeled Scabland in Washington and Idaho. The meltwaters overspilled into the Lake Superior and Huron Basins and, as the ice continued to retreat, a series of newly created lake overflow routes was used to transport water from Agassiz eastwards. On each occasion, Lake Agassiz alternately filled and drained with each emptying event causing a series of catastrophic floods. The floods in the Lake Superior Basin eventually ended at 8,500 BP when northward recession of the ice margin established a connection between Agassiz and glacial Lake Ojibway to the east. By 8,400 BP Lakes Agassiz and Ojibway may have become confluent (Dyke and Prest 1987) (see Chapter 5).

At this time, a series of remarkable environmental changes took place in North America and led to probably the largest superflood known on the planet. The water bodies of Lake Agassiz and Lake Ojibway bordered the southern ice sheet margin along 3,100 km of its length. Most probably the remainder of the Laurentide ice sheet was in an advanced state of decay with both the southern sections of the Keewatin and Hudson Lobes undergoing widespread stagnation as well as Labrador ice. Farther north, the rise of Holocene sea level exceeded the rate of glacio-isostatic uplift and resulted in marine incursion of the Tyrrell Sea into the Hudson Strait and into northern Hudson Bay.

Farther east, the waters of glacial Lake Ojibway, whose surface stood somewhere between 450 m and 600 m above present sea level, suddenly breached the Hudson and Labrador ice masses to the north and drained catastrophically into the Hudson Bay Lowland resulting in the lowering of the level of Lake Ojibway by at least 250 m (Dyke and Prest 1987). The area of Agassiz/Ojibway waters that drained northwards probably exceeded 700,000 km^2. It is difficult to estimate accurately the volume of water that must have drained from beneath the ice. The figure is most probably between 75,000 km^3 and 150,000 km^3 – thus representing one of the greatest floods that took place during the Quaternary (Table 8.3). The rate at which this water drained through the Hudson Strait into the North Atlantic is not known – this calculation being dependent on whether the waters 'leaked' through (and beneath) the ice or, as is more likely, if the ice barrier was suddenly breached by escaping lake waters.

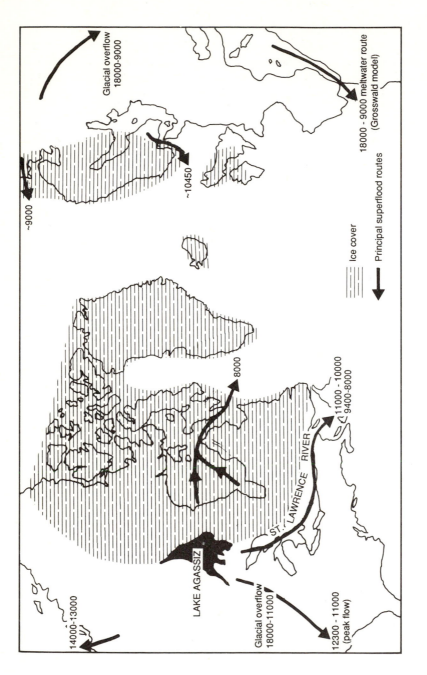

Figure 8.4 The principal routes of 'superfloods' that accompanied the melting of the last ice sheets. The ages of the respective flood events are approximate (based on various sources). Also shown are approximate positions of ice sheets at circa 11,000 BP.

Table 8.3 Approximate timing of major 'superfloods' during the melting of the last ice
   sheets. Crude estimates of flood volumes and equivalent values (cm) of global
   sea level rise are also given

Superflood	Date	Estimated water volume (km^3)		Equivalent global sea level (cm)
Glen Roy	10,500 BP (est.)	5		–
Baltic Ice Lake	10,450 BP	29,000	(100 m depth)	8
Lake Missoula	13,000 BP	2,000		0.55
Agassiz (via St Lawrence)	11,000 BP	23,000	(200 m depth)	6.4
Agassiz (via Hudson Strait	8,000 BP	75–150,000		19.4–41.6
Pur/Mensi (?)	9,000 BP	75,000	(50 m depth)	20.8
Pur/Mensi overflow	18–9,000 BP	48,000 m³s⁻¹		–
Agassiz overflow into Mississippi	13–11,000 BP	100,000 m³s⁻¹		–
Agassiz overflow via Lake Superior	9,400 BP	4,000		0.9

Eurasian superflood 'Events'

According to Grosswald (1980), superfloods were associated with the overflow
of meltwater from the large proglacial lakes that may have developed along the
southern margin of the Eurasian ice sheet. In this model, most of the glacial
meltwater that discharged from the southern section of the Eurasian ice sheet
eventually reached the eastern Mediterranean Sea the size of which was greatly
reduced due to glacio-eustatic lowering of sea level. Grosswald (1980) estimated
that the average annual discharge into the Caspian Sea during the Late Valdai
was at least 1,200–1,500 km³ per year, approximately equivalent to maximum
discharges of 48,000 m³ s⁻¹. The overflow of water mostly originated from the
Pur and Mensi lakes (Figure 4.1) that occupied 1,500,000 km² of the West Siberian Lowland. The Late Valdai history of overflow palaeohydrology is not known
in any detail (see Chapter 5), although it is thought that some of the largest rivers
had dimensions similar to those that overflowed from Lake Agassiz.

Inspection of Figure 4.1 indicates that the distribution and levels of the lakes
would have varied in response to fluctuations of the Late Valdai ice sheet margin.
As the ice retreated, it eventually resulted in the production of a new drainage
network that initially caused most meltwater to drain westwards into the
lowlands of northern Poland and Germany and ultimately into the North Sea.
The routes of meltwater flow are today indicated by a complex series of subpar-
allel interconnected trenches. As the ice sheet retreated northwards, individual
meltwater routes were successively abandoned in favour of parallel routes farther

Plate 9 Parallel roads of Glen Roy – shorelines of a former ice-dammed lake of Younger Dryas age in Glen Roy, Scottish Highlands. Photo J.J. Lowe.

north. Some of these channels were eroded adjacent to the ice sheet margin while others may have been produced by subglacial meltwater erosion – collectively they are known as *Urstromtaler* (Flint 1971).

An intriguing point also proposed by Grosswald is that regional deglaciation may have also resulted in the discharge of westward-flowing meltwater into the Baltic Ice Lake in southern Scandinavia. At a late stage during ice sheet deglaciation, ice sheet disintegration in the Ob–Yenisei lowlands resulted in a reversal of the southward and westward drainage routes to create a new pattern of northward flow into the Arctic Ocean.

In Scandinavia, the drainage of the Baltic Ice Lake during the Younger Dryas may have resulted in the occurrence of superfloods across parts of southern Sweden at the end of the Younger Dryas (Figure 5.3). Bjorck and Digerfeldt (1986) consider that the lake appears to have fallen in two stages with the first drainage event having taken place westwards at 11,250 BP and resulting in a lake level fall of 10–15 m. A second and much larger jökulhlaup took place near 10,450 BP when the lake level fell by 28 m. During this event, the ice barrier appears to have been breached by the lake waters near Mt Billingen in central Sweden. It is likely that the Billingen flood waters drained across the Kattegat/Skaggerak basin and across the land surface that then occupied the southern North Sea.

In the British Isles, the catastrophic drainage associated with the emptying of a glacier-dammed lake in Glen Roy in the Scottish Highlands is particularly well documented (Plate 9). The lake appears to have drained subglacially as a catastrophic jökulhlaup at the end of the Younger Dryas with floodwaters, estimated to have reached a peak discharge of $22,500\,m^3\,s^{-1}$, having emptied into Loch Ness where they caused an 8.5 m increase in lake level. Thereafter, a thick (up to 39 m) accumulation of gravel was deposited in the Inverness area as the floodwaters reached the sea (Sissons 1979b, 1981b). The catastrophic emptying of the Glen Roy lake is also thought to have been accompanied by active faulting and earthquake activity (Ringrose 1989). It is not known, however, whether an earthquake triggered lake drainage or if the emptying of the lake caused an earthquake.

SUMMARY

It is now widely known that the relationship between river terrace evolution and secular environmental changes is highly complex. Traditional interpretations relating river terrace development to changes in global climate (e.g. those of Penck and Bruckner) have been replaced by considerably more detailed models of terrace evolution in which attention has focused on local and regional palaeoenvironmental changes. However, as more is known about the nature of Late Quaternary environments, it has become clear that numerous high-magnitude fluvial 'events' have profoundly influenced the morphology and sedimentology of individual river systems. Indeed, in some instances the magnitude of these events may have been sufficient to have indirectly led to changes in global climate. It is also evident that the chronology of river discharge variations has differed markedly between middle and low latitudes.

In the mid latitude regions of the northern hemisphere during the last glaciation, many river systems were subject to greatly increased discharge due to the drainage of meltwater from the Eurasian, Laurentide and Cordilleran ice sheets. Thus, large volumes of meltwater drained southward from the Eurasian ice sheet into the Black Sea and eastern Mediterranean. The greatest change in the style of fluvial activity appears to have commenced soon after 12,500 BP and may have lasted until 11,000 BP. This period of climatic warming, characterised by rapid ice sheet thinning and retreat, may also have been associated with a major reorganisation in global climate from a glacial to an interglacial mode.

In the Mississippi drainage basin, the hydrological response to rapid ice melting at this time was extremely complex. This was largely due to the development of large proglacial lakes along the southern margin of the ice sheet. Greatly increased fluvial discharge occurred in the Mississippi–Missouri–Ohio drainage basin between 12,500 and 11,500 BP and was accompanied by increased sediment yield and the extensive deposition of braided river gravels.

Ice sheet melting in the northern hemisphere was also associated with the occurrence of several superfloods that took place as a result of the breaching of ice barriers by lake waters. The first group of these took place soon after 14,000 BP due to the (periodic) emptying of glacial Lake Missoula and the passage of floodwaters across the Channeled Scabland. A much larger flood took place at approximately 11,000 BP when Agassiz lake waters breached an ice barrier and flooded eastwards through the St Lawrence into the North Atlantic. Soon afterwards, near 10,450 BP, a large jökulhlaup of comparable magnitude drained from the Baltic Ice Lake into the North Sea. The largest flood, however, did not take place until 8,000 BP when the waters of glacial Lake Agassiz–Ojibway drained catastrophically through the Hudson Strait into the North Atlantic. The latter flood may have been the largest known to have taken place on Earth during the Quaternary. The final stages of ice sheet melting in the northern hemisphere between 9,000 and 7,500 BP were associated with the 'metamorphosis' of many rivers in the middle latitudes with the significant lowering of discharge leading to the widespread development of meandering channels.

In low latitude regions during the culmination of the last glaciation, many of the world's major rivers were greatly influenced by aridity, particularly between 25,000 and 18,000 BP. In both the Nile and Amazon Basins, river discharge was very low during this period while arid conditions are also indicated by marked increases in offshore aeolian sedimentation. Soon after 12,500 BP, however, the melting of the northern hemisphere ice sheets appears to have been accompanied by a rise in East African lake levels that, in the case of Lake Victoria, led to the overflow of lake waters into the Nile drainage basin and the occurrence of major flooding in the lower Nile. In West Africa, many rivers maintained relatively high seasonal discharges between 12,000 and 8,000 BP due to a displacement in the

160

position of the Intertropical Convergence Zone and more vigorous monsoonal circulation. A possible exception to this trend was during the Younger Dryas when both river discharge and lake levels were low.

The details described above present in approximate outline only the most important palaeohydrological changes that took place during the last glacial period. It is clear that some of the flood 'events' that took place were of phenomenal magnitude and, to date, have not been considered in detail by hydrologists. Broecker et al. (1989) have drawn attention to the important effects that some of the Laurentide ice sheet superfloods may have had on global climate. In doing so, they may have provided valuable data that may be used in future to develop numerical models of the hydrological response to the global environmental changes of the last ice age.

9

ICE AGE AEOLIAN ACTIVITY

INTRODUCTION

During the last glaciation vast areas of the Earth's surface were affected by pronounced aridity. As a result, wind action was a particularly effective and widespread agent of geomorphic change. This view contrasts with the old belief that 'pluvial' wet conditions prevailed during the last glacial maximum (see Chapter 7).

The increase in aeolian activity was due to the expansion of arid and semi-arid regions as a result of changes in atmospheric circulation that led to profound changes in the global distribution of rainfall. During this period, extensive dune fields were produced throughout large areas of Africa, India, Australia, North America and to a lesser extent in South America (Thomas 1989). In other mid latitude areas, great thicknesses of unstratified silt, known as loess, were deposited over wide areas. Most of the windblown loess sediments were derived from glacial outwash plains that bordered the middle latitude ice sheets. However, in some areas (e.g. China) the formation of loess may have originated in deserts due to the combined influence of frost and salt weathering. In many subtropical desert regions, the aeolian transport of loess also led to widespread dust deposition over ocean surfaces and eventual sediment deposition on the ocean floor.

The detailed studies that have been undertaken on loess stratigraphy in different parts of the world have also provided valuable information on the patterns of climatic change that took place during the Late Quaternary. Loess stratigraphy therefore provides one of the best terrestrial records of past climatic changes. In some areas, loess sediments are developed upon palaeosols and dating of these has enabled scientists to estimate the principal time periods of aeolian deposition. Since most loess appears to have been deposited during periods of cold climate, the dating of palaeosols provides an approximate record of global changes between warm and cold climate.

Plate 10 Thick loess sequences, Fen River valley area, Shanxi province, central China. A palaeosol can be seen as a dark horizon beneath the uppermost (Malan) loess deposits. Photo B. Percy-Smith.

LOESS

Origin

Loess may be defined as a wind-deposited silt, commonly accompanied by some clay and some fine sand. It consists of extremely well-sorted, fine-grained sediment and is mostly composed of silt-size quartz grains but it may also be highly calcareous. In general, most loess consists of particles ranging between 20 and 60 microns in diameter although considerable quantities of finer-sized material may also be present. Frequently loess accumulations are highly cohesive, having been consolidated due to the precipitation of secondary calcium carbonate. The individual particles are angular while the porosity of most loess is usually greater than 50 per cent. Erosion of loess by fluvial activity may often lead to the production of intensely gullied topography with steep, and in many cases almost vertical, cliffs.

Loess may form in two broadly distinct environments. First, it commonly originates as glacial rock flour in outwash plains during periods of continental ice sheet glaciation. By contrast, loess may also be produced in so-called 'warm' desert areas due to the combined action of frost and salt weathering (e.g. in Soviet Central Asia) (Pye 1987). In many areas loess sediments are extremely complex in their origins since they have been extensively reworked and redepo-

Plate 11 Series of vegetated relict Late Quaternary dunes in Gokwe District, west central Zimbabwe. Photo D. Thomas.

sited by colluvial and alluvial processes (Russell 1944; Pye 1984, 1987) (Plate 10).

Loess that originates in desert environments tends to be more poorly sorted than it is in periglacial environments. Most loess can be transported great distances beyond its source area, and is deposited in a quite different environment where vegetation may bind the sediment. Therefore, whereas cold climate loess originates in a periglacial environment, it is not necessarily indicative of such an environment in its area of deposition (Washburn 1979).

Most accumulations of cold climate loess are composite in age, the deposition of each stratigraphic unit having taken place during an individual period of glaciation. Frequently, individual loess units in a vertical profile are separated by palaeosols. In general, the thickness and grain size of loess increases exponentially towards the source area. Often, loess may contain large numbers of mollusca that may be used to indicate the former environment of sediment deposition (Leonard and Frye 1954). Leonard and Frye (1954) considered that loess deposition normally took place over moist vegetation surfaces where the sediment could be trapped easily.

Distribution

Loess deposits cover almost 10 per cent of the total land surface of the Earth while in the United States approximately 30 per cent of the landscape possesses a

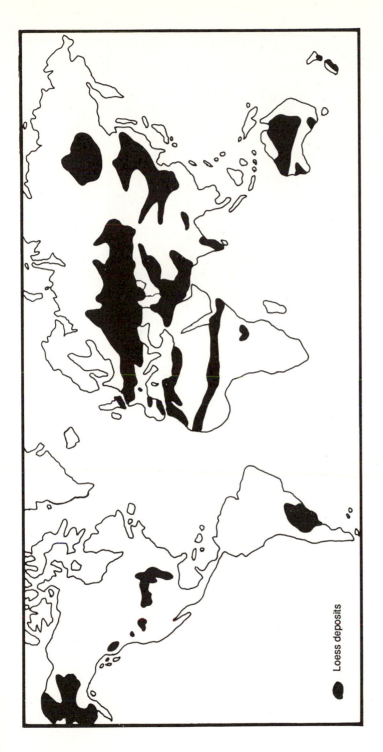

Figure 9.1 Global distribution of loess (after Catt 1988).

Loess deposits

Figure 9.2 Distribution of loess in Europe (after Flint 1971).

loess cover. The greatest thicknesses of loess occur in China where, locally, thicknesses of over 300m have been recorded (Figure 9.1). The global distribution of cold climate loess indicates that most of the sediment occurs outside the limits of the last ice sheets in the middle latitudes (Figure 9.2). However, it is not always easy to distinguish cold climate loess from that which has originated in continental interior deserts.

Europe

Loess deposited during the last glacial maximum is widespread throughout Europe. The deposits can be traced intermittently across northern France, Belgium and the Netherlands but the greatest accumulations occur in eastern Europe (Figure 9.2). The majority of European loess is derived from glacial outwash sediments and exhibits a progressive increase in extent and thickness from west to east. In northern France, the origin of loessic sediments has been related to a source in the English Channel during glacial periods when global sea level was low. By contrast, loess sediments in the Rhine valley, deposited by westerly winds, are thicker on the east of the river, particularly on west-facing slopes. Farther east, widespread accumulations of reworked loess occur near the largest rivers that transported large volumes of glacial meltwater and sediment during the last glaciation. Thus, thick loess accumulations (up to 50m) occur in the Danube, Don, Dneiper and Volga Basins.

North America

Loess of Late Wisconsin age is widespread in North America. Although it occurs over wide areas of eastern Washington, northeastern Oregon, southern Idaho and central Alaska (where it is up to 60m thick), the most extensive areas occur in mid continental United States (Figure 6.3). In this area it extends eastwards from the Rocky Mountains across the Great Plains and the Central Lowland into Pennsylvania (Flint 1971). Extensive areas of reworked loess occur in the Ohio–Missouri–Mississippi drainage basins as well as along the eastern flanks of the lower Mississippi valley. Throughout these areas, the thickest loess occurs near major river valleys while there is also a progressive thinning of loess downwind from the presumed source areas. A general characteristic of most loess in North America is that it was initially derived from glacial outwash during periods of glaciation and later reworked by fluvial and colluvial processes.

China

On the Chinese Loess Plateau, loess deposited during the last glacial maximum changes from a silty loess to a clayey loess in a downwind (southeasterly) direction as it becomes thinner. During isotope stage 2, a relatively thin covering of loess (usually between 10 and 30m) was deposited with an inferred average rate of deposition of 0.35 mm per year. The most recent view is that most of the Chinese loess is desert rather than glacial in origin (Pye 1984, 1987) (Figure 9.3). However, according to Smalley and Smalley (1983) most of the loess silts in Soviet Central Asia and in China were initially formed by glacial processes and frost weathering and thereafter transported to desert plains by rivers before being finally transported and deposited by wind (Plate 10).

Several factors appear to have been conducive to the widespread deposition of loess in China. One view is that loess deposition may have been favoured by increased continentality caused by a lowered sea level. It may also have been assisted by an increase in the number of cyclone-induced sandstorms in the Gobi Desert and to more effective easterly transport of dust by a westerly jet stream located north of the Tibetan Plateau and more strongly anchored in its position due to Late Quaternary tectonic uplift of this region (Tungsheng *et al.* 1986; Pye 1987). A major uncertainty, however, is the scale of glaciation in Tibet during the last glacial maximum. Thus, the extensive Tibetan ice sheet proposed by Kuhle (1987, 1988) would have provided large volumes of glacial rock flour as a source of cold climate loess in China. By contrast, the more restricted scale of Tibetan glaciation envisaged by Tungsheng *et al.* (1986) and Benxing (1989) implies that most Chinese loess is likely to be non-glacial in origin.

LATE QUATERNARY LOESS STRATIGRAPHY

China

Loess in China covers an area of at least 440,000 km^2 (Tungsheng *et al.* 1986) (Figure 9.3) and is thickest in the Loess Plateau of the middle Yellow River. Although most Chinese loess appears to have been deposited during the last 1 Myr, significant thicknesses (up to 130 m) of loess were deposited during the Late Quaternary (Kukla 1987). Some Chinese loess is derived from glacigenic sediments on the Tibetan Plateau and has been deposited downwind of this source area. However, very large quantities of desert loess also occur and are derived from arid areas to the north and west of the main loess belt. Tungsheng *et al.* (1986) have argued that in China there is a progressive NW to SE transition from rock desert to gravel desert (gobi), to sand desert, then to sandy loess, silty loess and fine loess.

The Chinese Late Quaternary loess unit is known as the Malan loess and is typically not more than 30 m in thickness. Its deposition appears to have been interrupted by periods of soil development, particularly during isotope stage 3. A palaeosol of last interglacial age also underlies the Malan loess. Comparison of the distribution of the Malan loess with that of older loess accumulations indicates that deposition was very widespread, reaching latitudes 25–30°N. This increase in the area of loess deposition has been considered indicative of the occurrence of a particularly dry and cold continental climate, perhaps caused by Late Quaternary uplift of the Qinghai-Xizang Plateau. In China, the principal

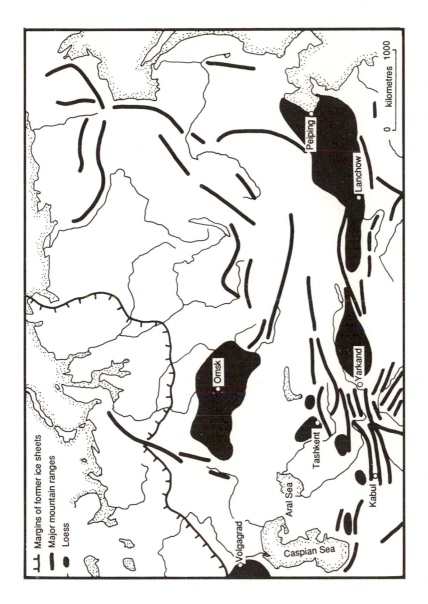

Figure 9.3 Distribution of principal loess areas in central and northern Asia (dark areas). The approximate margin of the Eurasian ice sheet and the location of major mountain ranges are also shown (after Flint 1971).

period of Malan loess deposition in northern China appears to have commenced near 30,000 BP and to have ended by 10,000 BP (Pye 1987).

Recently, much interest has focused on the observations of Kukla *et al.* (1988) who have noted a close correlation between the magnetic susceptibility of Chinese Malan loess sediments and the oxygen isotope record derived from ocean sediments (Figure 9.4). The results suggest that aeolian deposition in central China and global changes in ice volume during the Late Quaternary were both linked to and controlled by changes in the general circulation of the northern hemisphere (Kukla *et al.* 1988). The evidence provided by Kukla *et al.* shows that a major increase in loess deposition commenced at the end of isotope stage 5 – coinciding closely with the timing of rapid northern hemisphere glaciation inferred by Ruddiman *et al.* (1980).

Eastern Europe

Deposition of loess in Europe during the Late Quaternary was interrupted on occasions by periods of interstadial warmth and soil development. In European USSR, for example, loess deposition appears to have been interrupted during the Krutizk interstadial (between 68,000 and 58,000 BP) as well as perhaps also during isotope stage 3 between 28,000 and 25,000 BP (Velichko and Faustova 1986). Of these, the period of non-glaciation between 58,000 and 68,000 BP was marked by widespread soil development in the Don and Dneiper Basins. Thus, an early period of loess deposition in eastern Europe appears to have taken place prior to 68,000 BP but after the last interglacial, with a second period of loess deposition having taken place between 25,000 and 18,000 BP. The timing of loess deposition is broadly similar to that in China and tends to support the view that the major Late Quaternary glaciations were approximately synchronous across the northern hemisphere.

North America

The relative chronology of loess deposition in North America during the Late Quaternary is broadly analogous to that encountered in Europe and China although there is some uncertainty regarding the absolute ages of the oldest deposits. Loess units older than the Late Wisconsin are present in North America but they are not well dated. The lowest loess units are indirectly dated in some areas where they rest upon a palaeosol of last interglacial (Sangamon) age. In Iowa a lower loess may be as old as 31,000 BP while similar 'old' loessic sediments also occur in Illinois (Roxana silt) and Nebraska (Gilman Canyon loess) (Hallberg 1986). Johnson (1986) has argued, however, that loess deposition in Illinois may have commenced as early as between 50,000 and 45,000 BP. In Indiana, Iowa, Illinois and Nebraska, palaeosols dated between 29,000 and 25,000 BP occur beneath loess of Late Wisconsin age and demonstrate that glacial age aridity did not commence until after this time (Ruhe 1986).

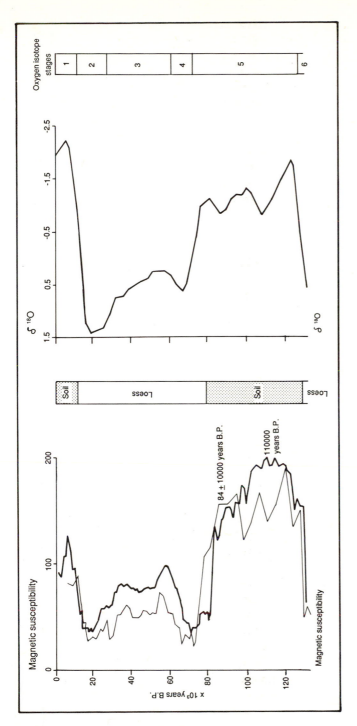

Figure 9.4 Profile of magnetic susceptibility of Chinese loess sequences compared with the oxygen isotope record for the Late Quaternary (after Kukla *et al.* 1988).

DUNES

Origin

Sand dunes form in both warm and cold deserts as well as along coastlines. The principal requirements for dune formation are an abundant sediment supply and an absence of a vegetation cover and sufficient wind to move sediment. In cold climate environments adjacent to glaciers and ice sheets, high velocity katabatic winds often produce a variety of dune forms.

It is difficult to use fossil dunes as palaeoenvironmental indicators since not only is it unclear how active dunes are formed by present-day winds but it is also difficult to distinguish active dunes from those that are fossil. Despite these uncertainties, dune orientations have been used to reconstruct regional patterns of former atmospheric circulation. Such analyses may be criticised since the prevailing winds may not be those that cause the dune building (cf. Thomas 1989). Therefore, dune orientations can only indicate general patterns of former wind direction.

Distribution of fossil dune fields

The dramatic shifts in the world's major vegetational zones during the last glaciation led to a corresponding expansion in the locations of the world's major tropical and subtropical deserts. Many of these deserts increased greatly in size as they responded to changes in temperature, windiness, precipitation and evaporation. Many of the dune fields produced during the last glaciation are now fossil (Figure 9.5) (Plate 11). Some have been eroded while others occur beneath actively forming dunes and are separated from these by fossil soils and, in some cases, archaeological remains. Other dune fields are now the sites of lakes. Sarnthein (1978) observed that, during the last glaciation, much of the land area between 30°N and 30°S was characterised by vast deserts located to the north and south of the Intertropical Convergence Zone. In this area, the extent of tropical rainforest and adjacent savanna was reduced to a narrow corridor, in places only a few degrees of latitude in width.

Considerable expansion of deserts took place in Africa (Heine 1982) (Figure 9.6). In southern Africa, there was an enormous increase in the area of the Kalahari (Grove 1969) across the present area of the Congo rainforest and into Botswana and Zimbabwe. In northern areas of Africa, dune fields produced during the last glaciation extended over 700 km farther south than they do at present (Grove 1969). Many of the fossil dune fields in this area seem to indicate that they were formed along the southern margin of an expanded Saharan subtropical anticyclone associated with prevailing northeasterly winds. In East Africa, the Qoz fossil dunes located both west and east of the White Nile indicate greatly increased aeolian activity during the last glacial maximum (Figure 9.6). These conclusions, however, should be tempered with caution since many of the

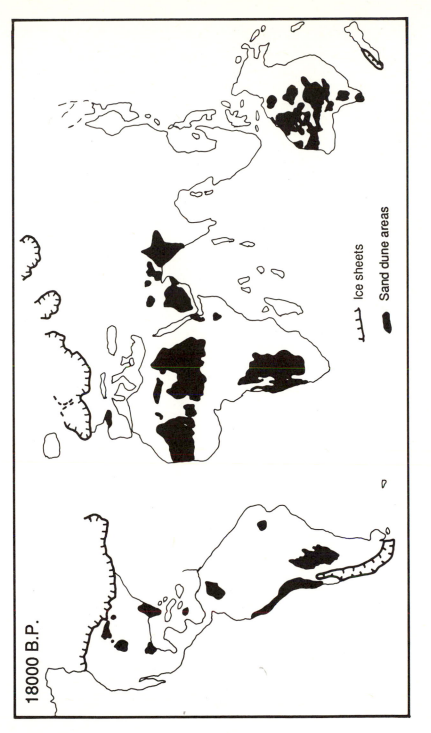

Figure 9.5 The distribution of active sand dunes (ergs) during the last glacial maximum (after Goudie 1983).

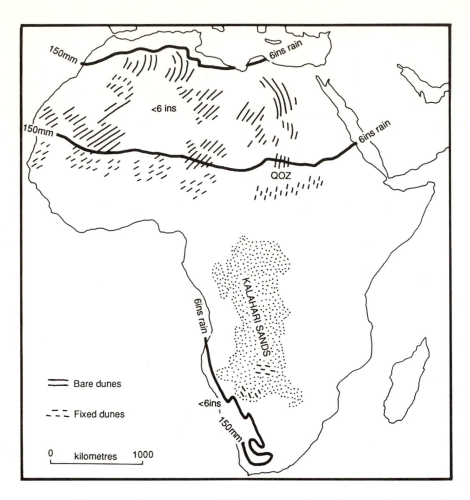

Figure 9.6 The past and present extent of blown sand in Africa (after Goudie 1983). The present annual rainfall range is also shown.

fossil dunes are not well dated due to the lack of organic matter and it is possible that some dune fields may be of different ages (Thomas and Goudie 1984).

In western India and eastern Pakistan very extensive areas of desert developed during the last glaciation (Verstappen 1970; Goudie *et al.* 1973) (Figure 9.5). The great increase in aridity across the Indian subcontinent during this period was largely due to the creation of permanent high pressure over the Eurasian continent that prevented the invasion of monsoonal air masses into the subcontinent.

In South America, fossil dune fields produced during the last glaciation have been observed in parts of the Orinoco and Amazon drainage basins. Extensive dune fields also occur in the Pampas of Argentina (Tricart 1975; Goudie 1983) (Figure 9.5). Although patterns of aeolian deposition in the Pampas may have been profoundly influenced by neighbouring Patagonian ice masses, the dune fields of the Orinoco and Amazon may have resulted from changes in the position of the January NE monsoon (Bigarella and Andrade 1965; Tricart 1975). The most likely cause of this change is that the Intertropical Convergence Zone was displaced northwards in response to strengthened subtropical anticyclones in the eastern Pacific and South Atlantic Oceans.

The Australian continent was also profoundly influenced by arid conditions and dune building during the last glacial maximum (Figure 9.5). Bowler (1978) has demonstrated that the development of many Australian dune fields commenced soon after 25,000 BP and coincided with the disappearance of many

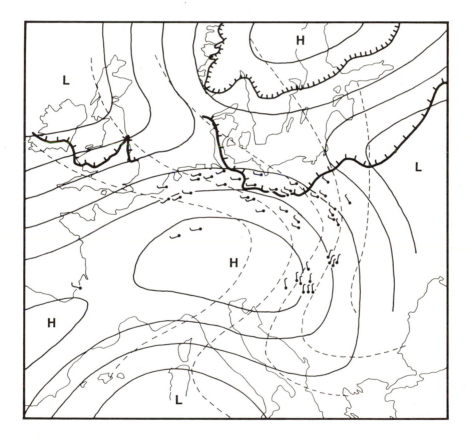

Figure 9.7 Inferred summer atmospheric pressure and wind directions in Europe during the Lateglacial period. Ice margins are shown for the beginning and end of Lateglacial time. The principal cells of high (H) and low (L) pressure are also shown (after Washburn 1979).

lakes. The orientations of the principal dune fields exhibit a characteristic anti-clockwise pattern associated with a strong anticyclone located over central Australia. The anticyclone was sufficiently enlarged that it provided conditions suitable for dune development over most areas of southern Australia as well as northern Tasmania. The extensive areas of glacial-age dunes may also be related to a decreased supply of moisture (due to the strengthened anticyclone) as well as reduced temperatures, a reduced vegetation cover and increased continentality due to a lowered sea level.

Extensive dune fields also formed in Europe during the last glacial maximum although these were mostly produced due to the combined influence of severe periglacial conditions and intense wind activity in areas located south of the last Eurasian ice sheet. Reiter and Poser (cited in Washburn 1979) used dune orientation data for Europe to reconstruct patterns of atmospheric circulation for the Lateglacial period (Figure 9.7). The inferred palaeowind reconstructions for Europe provide a fascinating insight into former atmospheric circulation (Goosens 1988). They show that, whereas the dune fields were deposited in association with an anticyclone over central and northern Europe, the loess sheets that increase in thickness eastwards must have been deposited by winds related to a permanent anticyclone over the Eurasian ice sheet to the north.

EVIDENCE FROM OCEAN AND ICE CORES

Ocean cores

A number of authors have drawn attention to evidence from ocean cores that indicates an increase in aeolian sedimentation during the last glacial maximum. For example, Damuth and Fairbridge (1970) observed that cores in the Atlantic off Brazil contained relatively high proportions of feldspar minerals and less quartz grains when compared with those of the Holocene. They considered that this provided evidence of decreased chemical weathering in the Amazon basin during the last glacial maximum, possibly caused by greater aridity and aeolian activity. Similarly, Bonatti and Gartner (1973) concluded from studies of the proportion of quartz grains in several Caribbean cores that the neighbouring continental land areas must have been characterised by increased aridity during the last glaciation.

Diester-Haas (1976, 1980) and Parkin and Shackleton (1973) have shown from ocean cores off northwest Africa that considerably increased quantities of Saharan dust were deposited upon the ocean surface during glacial periods. This not only indicates the increased vigour of subtropical anticyclonic circulation over Saharan Africa at this time but it also implies that the frequency of offshore winds may have increased, perhaps assisted by more pronounced oceanic upwelling.

Ice cores

Recently, a detailed record of Late Quaternary dust deposition has been provided from the Vostok ice core in Antarctica (Petit *et al.* 1990) (Figure 9.8). The data show that there was a significant increase in the rate of aeolian deposition during oxygen isotope stage 2 but that this increase may have commenced as early as 30,000 BP. However, a marked increase in the rate of deposition also took place

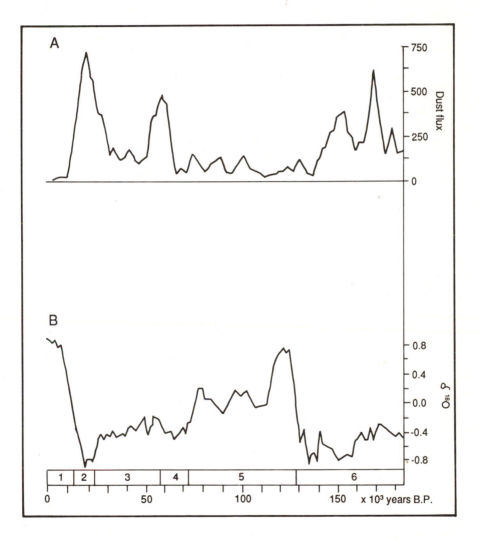

Figure 9.8 The smoothed Vostok dust flux record (A) and the oxygen isotope record from Vostok (B) showing maximum rates of dust deposition during isotope stages 2 and 4 (after Petit *et al.* 1990).

during the transition between isotope stages 5 and 4. During the remainder of the Late Quaternary, rates of aeolian sedimentation appear to have been low.

SUMMARY

It is often difficult to interpret the palaeoclimatic significance of Late Quaternary aeolian sediments and landforms. Although the chronology of aeolian deposition during the Late Quaternary is greatly dependent on accurate radiometric dating, very severe problems are encountered when attempts are made to reconstruct former climates using aeolian sediments and landforms of known ages. For example, it is difficult to use fossil dune fields as indicators of former wind direction since not only is it unclear how active dunes are formed by present-day winds but it is also difficult to distinguish active dunes from those that are fossil. Nevertheless, where there is reason to believe that particular fossil dune alignments of a given age are indicative of specific wind directions, it may prove possible to evaluate patterns of atmospheric motion predicted by GCM computer simulations (see Chapter 3). Similarly, it is often very difficult to differentiate loess that is glacial in origin from that which originates in deserts.

Given these limitations, there is little doubt that there was a pronounced increase in aeolian activity during periods of Late Quaternary glaciation. This is very clearly illustrated by Kukla *et al.* (1988) who have shown that there is a strong correlation between Late Quaternary loess deposition in central China and global changes in ice volume inferred from oxygen isotope stratigraphy. In China, there was a major increase in the rate of loess deposition at the end of oxygen isotope stage 5. A similar pattern is evident for the Soviet Union where a major period of loess deposition is considered to have taken place after the last interglacial yet prior to 68,000 BP.

Although the history of loess deposition during isotope stages 3 and 2 is unclear, there is some agreement that loess deposition resumed sometime after 30,000 BP and probably nearer 25,000 BP. Aeolian activity was widespread in North America and eastern Europe until 18,000 BP and may have continued in northern China until nearer 10,000 BP. The time periods during which there was a hiatus in loess deposition are also instructive since they may be indicative of reduced glacial activity and a milder climate. The data from eastern Europe imply that one such period occurred between 68,000 and 58,000 BP while a second and much later interval took place in eastern Europe and North America between 29,000 and 25,000 BP.

During the last glacial maximum, increased aridity was associated with widespread aeolian activity. One of the most marked changes in low latitude areas was an immense expansion of the world's tropical and subtropical sand deserts that led to the widespread development of dune fields. In southern Africa there was an enormous increase in the area of the Kalahari while, in northern Africa, dune fields extended nearly 700 km farther south than they do at present. In South America, fossil dune fields of a similar age were formed in parts of the

Orinoco and Amazon drainage basins. Dune fields of last glacial age are also known to have developed in India, Pakistan and Australia. There is therefore great potential to use the data on Late Quaternary loess and dune field distributions to improve the calibration of the general circulation models of ice age atmospheres that have so far been developed.

10

LATE QUATERNARY VOLCANIC ACTIVITY

INTRODUCTION

In recent years, the causes of ice ages have been increasingly attributed to mechanisms related to Milankovitch orbital processes (see Chapter 13). These have been accompanied by a reduced interest in the influence of volcanic activity on climate cooling. Yet it is well known that recent volcanism, particularly at low latitudes, can result and has resulted in relatively brief (decadal) intervals of global climatic cooling (Lamb 1982). However, there has always been uncertainty about whether or not extremely large eruptions in the past have been capable of inducing global cooling sufficient to cause glacier and ice sheet expansion (Flint 1971). Although there are no easy answers to such questions, a first step is to discover the magnitude and frequency of the largest volcanic eruptions that took place during the Late Quaternary. Was it the case, for example, that the largest Late Quaternary volcanic eruptions were of magnitudes greater than those of the Holocene (e.g. Mt Mazama, Santorini, Tambora) – and did they have more significant impacts on climate? Additionally, it is important to discover if the growth and decay of Late Quaternary ice sheets and global changes in sea level may actually have triggered episodes of volcanic activity as some have suggested (Hall 1982; Sejrup *et al.* 1989).

Individual eruptions also result in the eventual deposition of air-fall tephra on the floors of the world's oceans and on continents where they provide valuable stratigraphic marker horizons that, in turn, permit correlation between oceanic and terrestrial sediment sequences. Owing to their time-synchronous deposition over wide areas, individual tephra layers are therefore of considerable value in the reconstruction of Late Quaternary environmental changes. For example, the presence of these time-synchronous marker horizons in ocean sediments has enabled scientists to date former changes in ocean temperature with very great accuracy. Similarly, periods of increased volcanic activity are often associated with peaks of acidity within ice cores that can sometimes be related to particular eruptions.

VOLCANOES AND CLIMATE

Processes

Explosive volcanic eruptions are commonly acidic in character and are associated with the release into the atmosphere of fine-grained volcanic ash and sulphitic aerosol compounds. The vertical column height of ash in the atmosphere not only depends on the magnitude of the eruption but is also a function of the rate of magma discharge from the vent. If a volcanic eruption is confined to the emission of ash into the troposphere (e.g. Mt St Helens 1980), much of the ash tends to be rapidly washed out of the atmosphere via precipitation. However, if the emission results in the introduction of ash through the tropopause and into the stratosphere, the volcanic eruption may be accompanied by the development of a stratospheric dust veil. Since the average diameter of ash particles is typically 0.5 microns and the range in wavelength of the visible spectrum is between 0.3 and 0.8 microns, the stratospheric ash tends to absorb and scatter incoming solar radiation. It is believed that such decreases may result in global cooling. Lamb (1982) estimated, for example, that major volcanic eruptions in Japan and Iceland during 1783 may have caused an overall cooling of the northern hemisphere by 1.3°C but that this temperature lowering may have ceased after one or two decades. Similarly, the eruption in 1883 of Krakatau in the East Indies appears to have resulted in a slight reduction of global temperatures during the ensuing two decades.

The effect of explosive volcanic eruptions on climate not only depends upon their magnitude but is also related to the latitude at which the eruptions take place. For example, eruptions in high latitudes tend to produce stratospheric dust veils only over the hemisphere in which the eruption occurs. However, stratospheric dust over high latitude regions, due to the low angles of incident solar radiation, may be particularly effective in backscattering radiation and thus increasing the tropospheric cooling caused by the presence of ash in the atmosphere. By contrast, low latitude eruptions tend to result in the production of a stratospheric dust veil over both hemispheres due to poleward convection of air within Hadley cells both north and south of the Equator.

The length of time during which volcanic ash from a major eruption remains in the stratosphere is also an important factor in determining the influence of any particular eruption on global climate. In general, tephra produced by a major eruption will take several days to encircle the Earth and approximately 6 months to produce a global veil. The ash particles settle out of suspension from the atmosphere according to their individual densities but, in general, most ash will have been deposited after a year although the completion of ash deposition may not have taken place until approximately a decade after the eruption.

Certain high latitude eruptions may also produce tephra that is deposited upon sea ice or upon ice shelves. Tephra deposits that initially settle upon sea ice and are later deposited on the sea floor are often quite distinctive since they commonly occur in association with terrigenous debris. Rafting by floating ice may also result in the occurrence of quite coarse-grained fall-out tephra at great distances from the source volcano. Tephra that accumulates on the sea floor is substantially affected by the burrowing of bottom-dwelling organisms (bioturbation) as well as by dispersal and redeposition due to turbidity current activity. Bioturbation is of particular importance since it may cause mixing within relatively thick accumulations (up to 0.4 m) of marine sediment (Ruddiman and Glover 1972).

Tephra deposition may also have taken place upon the surfaces of the northern hemisphere ice sheets during the Late Quaternary although any evidence for such activity has since been removed during deglaciation. Volcanic ash of Late Quaternary age has, however, been identified in ice cores recovered from the Greenland and Antarctic ice sheets (e.g. Gow and Williamson 1971; Hammer *et al.* 1981). The occurrence of tephra in ice cores also often coincides with an increase in ice acidity due to the precipitation of sulphitic aerosol compounds.

Identification and dating of tephra layers

Tephra layers in terrestrial and ocean sediment sequences can be identified in several ways. These include the use of refractive indices, the mineralogy of the shards and studies of their geochemical composition using electron microprobe techniques (Fisher and Schmincke 1984). The results of electron microprobe

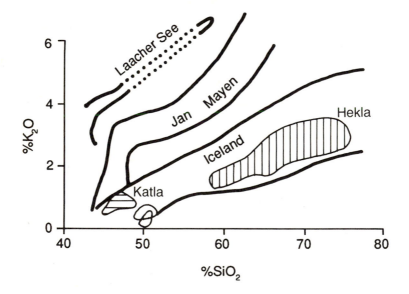

Figure 10.1 Geochemistry of several tephra deposits showing discrete plots of abundance of K_2O–SiO_2 (after Dugmore 1989).

analyses of volcanic shards can be matched against the geochemical composition of individual volcanic systems. It is particularly difficult, however, to trace individual tephra layers to the source volcano. In this respect, for example, the measured abundances of $K_2O–SiO_2$ of particular volcanoes can be compared with those of individual tephra layers (Figure 10.1). Often, the distribution of ash fall deposits provides important information since they are deposited downwind of the eruptive source (Mangerud et al. 1984).

In ocean sediments the relative dating of individual tephra layers is accomplished by a variety of techniques that include the cross-correlation of tephra layers with faunal assemblages in cores as well as the widespread use of oxygen isotope stratigraphy. The finite dating of some Late Quaternary tephra layers has been determined by K–Ar dating although this technique is usually only applicable to sediments older than 50,000 years. By contrast, the ages of younger tephra layers in marine sediments have been indirectly determined by calculating the ^{14}C ages of marine shells occurring above and beneath individual ash layers. On land, tephra layers are often well preserved within peat deposits as well as within lake sediments (Thompson and Bradshaw 1986; Dugmore 1989). The ages of tephra layers within peat are determined by the ^{14}C dating of adjacent peats. However, many tephra layers that occur within peat are not visible to the naked eye (Persson 1971, Dugmore 1989). Under these circumstances the identification of tephra layers may be accomplished using X-ray and palaeomagnetic techniques (Dugmore 1989).

MAJOR VOLCANIC ERUPTIONS AND LATE QUATERNARY CLIMATE

Introduction

Numerous ash layers produced from subaerial volcanic eruptions have been identified in ocean floor sediments throughout the world. Clearly, the largest explosive volcanic eruptions (and their associated tephra layers) are those most likely to have caused major global cooling. In this section, therefore, an attempt is made to focus on the largest volcanic eruptions during the Late Quaternary. The list is not exhaustive but the eruptions described are the largest, and are therefore of particular significance not only in the context of Late Quaternary climates but also in the wider aspects of stratigraphic correlation.

Early eruptions

Central American eruptions

Tephra horizons identified from numerous ocean cores from the Gulf of Mexico and the eastern equatorial Pacific indicate that four major volcanic eruptions took place in Mexico and Guatemala during the Late Quaternary (Rabek et al.

Plate 12 Part of lake-filled caldera, Toba, northern Sumatra. Photo M. Acreman.

1985). The ages of the Mexican eruptions have been estimated respectively to 110,000, 65,000 and 30,000 BP. The climatic significance of the eruptions is not known although it appears that the 65,000 BP eruption was the most extensive with the occurrence of tephra in ocean cores up to 400 km from the likely eruptive source.

The Atitlan eruption in the Guatemalan Highlands

The most widespread tephra horizon in Central America was first identified by Worzel (1959) from numerous ocean cores from the eastern Pacific where it occurs across an area of $1.25 \times 10^6 km^2$. The same ash layer was reported from cores in the North Atlantic and Gulf of Mexico (Ewing *et al.* 1958). Later, Hahn *et al.* (1979) described pyroclastic flow deposits from the Atitlan eruption in Guatemala whose trace element geochemistry was very similar in composition to the ash layer recognised in the ocean cores. This ash layer, known as the Los Chocoyos Ash, has now been found not only in the western Gulf of Mexico but also from the eastern Pacific Ocean and in Caribbean cores up to 1,600 km downwind of the eruption and across an ocean area of $6 \times 10^6 km^2$ (Bowles *et al.* 1973; Drexler *et al.* 1980) (Figure 10.2). The related volcanic source appears to have been the Lake Atitlan area of the Guatemalan Highlands (Hahn *et al.* 1979; Rose *et al.* 1979; Rabek *et al.* 1985).

184

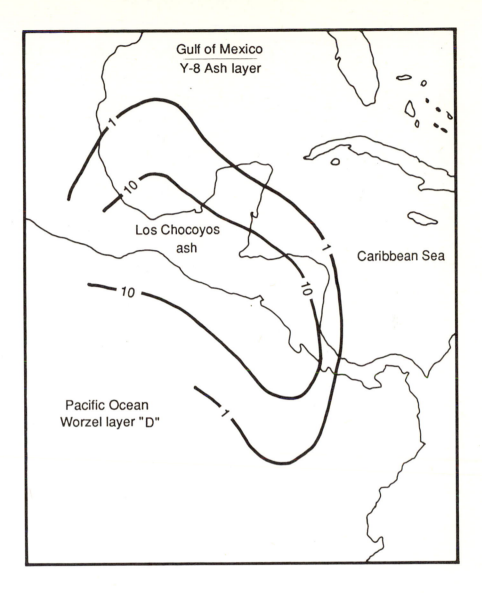

Figure 10.2 Contour map showing distribution and thickness (cm) of Los Chocoyos ash (after Drexler *et al.* 1980).

Kennett and Huddlestun (1972) concluded on the basis of planktonic foraminiferal studies from several Gulf of Mexico cores that an abrupt faunal change and hence climatic change took place in the region at 90,000 BP. They observed that this change coincided approximately with the widespread

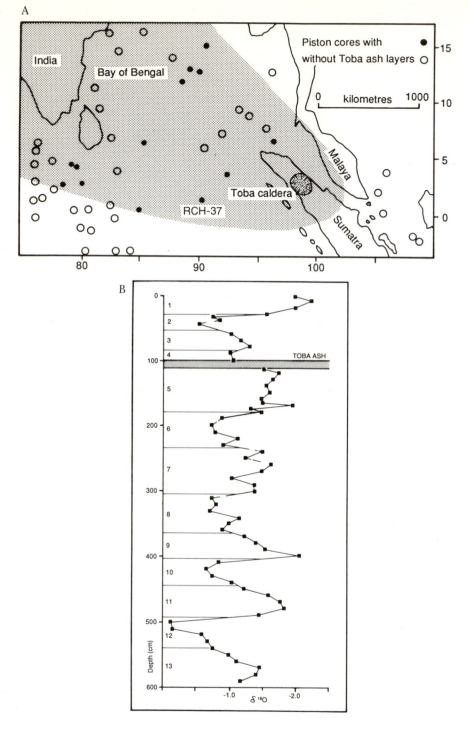

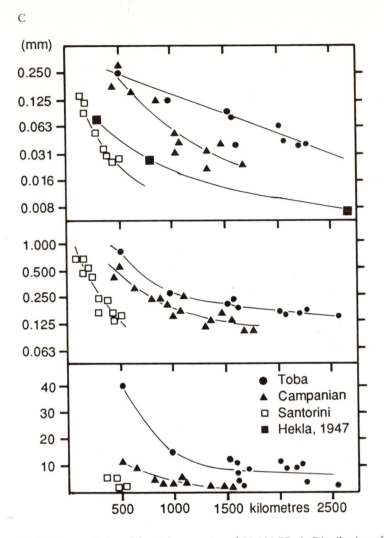

Figure 10.3 Characteristics of the Toba eruption of 75,000 BP. A. Distribution of tephra plume and occurrence of Toba tephra in deep sea cores. B. Oxygen isotope curve of core RC14-37 from the Indian Ocean showing stratigraphic position of Toba tephra. C. (Upper diagram) Plot of median grain size and distance for submarine fallout layers of Toba, Campanian and Santorini layers compared with the 1947 Hekla fallout. (Middle diagram) Plot of maximum particle size with distance. (Lower diagram) Plot of tephra thickness with distance (after Ninkovich *et al.* 1978; Fisher and Schminke 1984).

deposition of volcanic ash. The ash that they described has since been estimated to have been deposited at 84,000 BP (Drexler *et al.* 1980). Kennett and Huddlestun (1972) noted that the oxygen isotope stratigraphy for this time period that occurs in the Greenland Camp Century ice core shows a pronounced 'spike' indicative of global ice accumulation near this time (Johnsen *et al.* 1972) (Figure 2.6). They proposed that this period of rapid global ice build-up, which coincides with oxygen isotope substage 5b, was perhaps related to climatic cooling triggered by the Atitlan eruption. However, they also noted that the cause–effect relationship was difficult to resolve since, in the cores that they examined, tephra deposition appears to have commenced shortly after the initial cooling.

The Toba eruption in northern Sumatra

The largest magnitude explosive eruption that took place during the Late Quaternary was associated with the eruption of Toba, northern Sumatra (Figure 10.3) (Plate 12). This eruption dwarfs, both in scale and magnitude, all other volcanic eruptions during the Quaternary and may have profoundly influenced global climate. The early accounts of the eruption were provided by Van Bemmelen (1949) while more recent information has been provided by Ninkovich *et al.* (1978) and Chesner *et al.* (1990) who concluded that the Toba eruption was an order of magnitude larger than any other documented for the Quaternary. Ninkovich *et al.* (1978) noted that the Toba caldera has an area of approximately 3,000 km^2 and is surrounded by rhyolitic tuffs that occupy an area of 20,000–30,000 km^2 and which are several hundred metres in thickness. In neighbouring Malaya at Tampan, 400 km to the NE of the caldera, the ash layer deposited by the eruption is up to 1 m in thickness (Stauffer *et al.* 1980). The Toba ash layer has also been identified in numerous ocean cores located up to 2,500 km west of northern Sumatra (Figure 10.3) and on land in India (Rose and Chesner 1987). The associated tuffs have been K–Ar dated by Ninkovich *et al.* (1978) and $^{40}AR/^{39}Ar$ dated by Chesner *et al.* (1990) who concluded that the Toba eruption took place at approximately 75,000 BP.

Theoretical calculations of the volume of tephra deposited from the Toba eruption suggest that the dense rock equivalent volume of rhyolitic magma may have been as high as 2,800 km^3 (Rose and Chesner 1987). Fisher and Schmincke (1984) have estimated that if the Toba magma was erupted in approximately 10 days, the rate of magma discharge would have been approximately 10 million m^3 s^{-1}. Wilson *et al.* (1978) and Settle (1978) have, in turn, argued that the eruption column height in the atmosphere is closely related to the rate of magma discharge from the volcano. If this is the case, one may envisage that the column height of Toba tephra may have reached an altitude of 50–80 km (Ninkovich *et al.* 1978; Fisher and Schmincke 1984). Based upon shard morphology, settling velocities, shard distribution and an implied co-ignimbrite origin, Rose and Chesner (pers. comm.) suggest that this estimate may be too high by as much as a factor of five.

At present, the base of the stratosphere in low latitudes occurs at approximately 12 km while farther aloft, at 45–50 km, the stratopause separates the upper stratosphere from the overlying mesosphere. Consequently, it is not yet proven if the Toba eruption may have produced a tephra column that actually penetrated through the stratosphere and into the mesosphere. The presence of the Toba ash layer in several ocean cores permitted the cross-correlation of the K–Ar derived age with that determined by oxygen isotope stratigraphy (Ninkovich *et al.* 1978). The timing of the eruption appears to have occurred at the oxygen isotope stage 5/4 boundary, when, according to Ruddiman *et al.* (1980), there was a period of rapid and widespread ice sheet growth in the northern hemisphere. It is tempting to speculate that rapid northern hemisphere glaciation at 75,000 BP was triggered by the Toba eruption that provided a critical threshold for the initiation of widespread glaciation at a time when, according to the Milankovitch reconstructions, there was also a marked decline in summer solar radiation (Berger 1979). The possibility that the Toba eruption contributed to the initiation of an ice age is, however, dependent upon future research that may provide more information on the scale and timing of the eruption.

Antarctic eruptions

In an ice core from Byrd Station, Antarctica, Gow and Williamson (1971) noted evidence of considerable tephra deposition between 30,000 and 16,000 BP with peak ash deposition between 20,000 and 16,000 BP (Figure 10.4). The ash deposition appears to have been related to volcanic eruptions in Marie Byrd Land, Antarctica. Gow and Williamson (1971) estimated, by comparison with the isotope record from the ice core, that temperatures over the Antarctic troposphere during the last glacial maximum were reduced by as much as 2–3°C from their values at 30,000 BP. It is not possible to deduce the climatic effect of such widespread tephra deposition. Gow and Williamson, however, noted that since the global tropospheric cooling did not end until ash deposition had effectively ceased, it was possible that this cooling was caused by the presence of tephra in the Antarctic stratosphere.

North Atlantic Ash Zone 2

In the North Atlantic sea floor sediments there are only two major tephra horizons in the Late Quaternary sediments. The older of these was originally thought to have been deposited at 65,000 BP during the isotope stage 4/3 transition (Ruddiman 1977). However, more recent estimates point to an eruption nearer 57,500 BP (Ruddiman *et al.* 1980). The ash layer is known as North Atlantic Ash Zone 2 (NAZ2) (Ruddiman *et al.* 1980) and has an estimated ash volume of 7.6 km³ (Fisher and Schmincke 1984). However, the source of the ash is not known although it is likely to have been either Iceland or Jan Mayen (Bramlette and Bradley 1941; Ruddiman and Glover 1972).

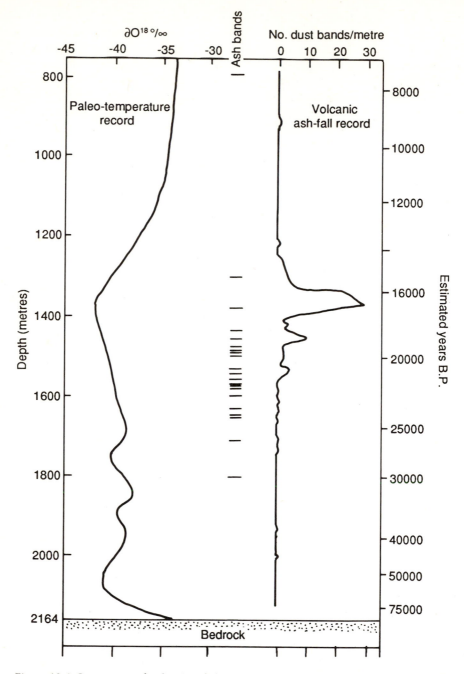

Figure 10.4 Occurrence of volcanic ash layers in Byrd Station ice core, Antarctica, and inferred palaeotemperature record. The sources of the respective ash layers are not known (after Gow and Williamson 1971).

Other major eruptions

Additional major eruptions and related phases of tephra deposition are known for the Late Quaternary but their climatological significance is uncertain. Among these is the deposition of the Old Crow tephra across north-central Alaska and the western Yukon (Westgate *et al.* 1983). In this region, the associated ash layer has been traced over a distance of 600 km. Its age is not yet known with certainty although it is thought to have been deposited between 105,000 and 87,000 BP (Schweger and Matthews 1985). The most widespread Late Quaternary ash in deep sea sediments from the eastern Mediterranean is the Y-5 ash layer (Thunell *et al.* 1979) that appears to have originated from the Campanian volcanic province near Naples. The age of the deposit is between 25,000 and 30,000 BP with an estimated volume of 100 km³ (Cramp *et al.* 1989). The Y-5 ash layer, by virtue of the great volume of erupted debris, therefore merits serious consideration as a trigger mechanism for global climate cooling but, as yet, insufficient information is available.

Later eruptions

North Atlantic Ash Zone 1

North Atlantic Ash Zone (NAZ1) is the most extensive ash layer contained within ocean sediments in the North Atlantic. The ash layer occurs in most North Atlantic ocean cores between 45 and 65°N and has an estimated ash volume of 1 km³ (Figure 10.5). Although it is generally agreed that the small ash volume (1 km³) is indicative of an eruption unlikely to have led to global cooling, the ash layer is of great importance due to its value as a time-synchronous marker horizon in North Atlantic marine and continental sediment sequences. The layer was initially dated to 9,300 BP (Ruddiman and McIntyre 1973; Ruddiman and Glover 1972, 1975) although Ruddiman and Glover (1975) observed that the ash layer was dispersed in some cores due largely to the effects of bioturbation. More recently, a distinct ash layer (the Vedde Ash Bed) has been discovered in western Norway within lake and marine sediments of Younger Dryas age (Mangerud *et al.* 1984) and dated by these authors at approximately 10,600 BP. In addition, Long and Morton (1987) have identified the Vedde ash layer in cores from both the North Sea and the northern North Atlantic (Plate 13). Mangerud and Long both noted that the Vedde ash layer is geochemically very similar to part of North Atlantic Ash Zone 1 identified by Ruddiman and McIntyre.

It has been concluded that the source volcano was most probably Mt Katla in southern Iceland that may have erupted explosively during this period from beneath a Younger Dryas ice cap (Long and Morton 1987). Hence the NAZ1 tephra layer may have been deposited across the North Atlantic Ocean and also upon parts of continental NW Europe and Scandinavia at 10,600 BP and not at 9,300 BP as previously believed. An additional implication is that widespread

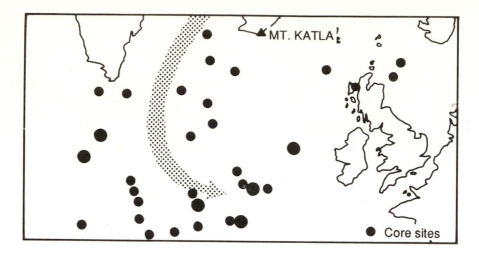

Figure 10.5 Location of cores known to contain North Atlantic Ash Zone 1 (the Vedde ash layer). The arrow shows the axis of the prevailing counterclockwise ocean gyre. Some of the ash may have been initially deposited on sea ice transported in this gyre and later deposited after ice melting. The data are based on Ruddiman and Glover (1975) and Long and Morton (1987).

tephra deposition took place upon the Fennoscandian (and possibly also the Scottish) ice sheet during the Younger Dryas.

One problem associated with the identification of North Atlantic Ash Zone 1 is that the ash may represent the product of more than one Icelandic volcanic eruption (Mangerud *et al.* 1984; Long and Morton 1987). For example, a prominent tephra layer with a probable Icelandic source (Grimsvatn), dated at 9,100 BP, is present in peat cores and lake sediments from the Faeroe Islands and in nearby ocean cores (Johansen 1975; Dugmore, pers. comm.). Marine tephra deposited by this eruption may also be present in North Atlantic ocean floor sediments but owing to the effects of bioturbation may be mixed together with the Vedde ash and North Atlantic Ash Zone 1 complex (e.g. Sejrup *et al.* 1989).

North Atlantic Ash Zone 1 has proved to be a very valuable stratigraphic marker horizon in ocean sediments and has been used extensively in the reconstructions of the Late Quaternary polar oceanic front fluctuations for the North Atlantic region (Ruddiman and McIntyre 1981a) (Figure 5.4). One of the most interesting aspects of the environmental changes deduced by Ruddiman and co-workers is the proposed sequence of Lateglacial environmental changes in the North Atlantic Ocean and NW European continent, whereby the occurrence of a brief yet severe period of cold climate during the Younger Dryas (see Chapter 5) throughout NW Europe was associated with a southward excursion of the polar oceanic front in the North Atlantic. Present understanding of this stadial period is greatly influenced by the model of Lateglacial oceanographic changes in the

Plate 13 Scanning electron micrograph of Vedde ash grains of Younger Dryas age, Alesund, western Norway. Horizontal bar shows scale in microns. Photo A. Newton and J. Finlay.

North Atlantic proposed by Ruddiman and McIntyre (1981b). The timing of these Lateglacial environmental changes is, however, largely dependent on acceptance of a 9,300 BP age for the NAZ1 tephra. It is possible to argue, on the basis of correlation with the Vedde ash, that the NAZ1 tephra was deposited at 10,600 BP and that the Ruddiman and McIntyre (1981a) model of northern hemisphere ice sheet deglaciation may be in need of reassessment. These interpretations should be treated with caution, however, since very few sites in which NAZ1 is present have so far been analysed geochemically.

Laacher See eruptions

Additional explosive volcanism took place in northern Europe during the Lateglacial. In particular, a series of Plinian eruptions took place at the Laacher See volcano in northern Germany at 11,000 BP (Bogaard and Schmincke 1985) (Figure 10.6). The eruption produced at least 5 km³ of magma (dense rock equivalent) and led to the deposition of tephra more than 50 m thick close to the eruptive centre (Bogaard and Schmincke 1985). Three separate plumes of tephra have been identified for this eruption of which deposition to the NE appears to have been the most extensive where it has been recognised up to 1,100 km from the eruptive source (Fisher and Schmincke 1984). In these eruptions, tephra was deposited over 170,000 km² of the northern European lowlands and also across southern Scandinavia. The Laacher See ash has been identified in over 120 Lateglacial sedimentary sections across Europe where it occurs within deposits of younger Alleröd age. The climatic consequences of the eruptions, however, are not known. The eruptions appear to have taken place near the beginning of the Younger Dryas cooling episode. However, it has not been demonstrated that there is a connection between these eruptions and Younger Dryas climatic deterioration.

Glacier Peak eruptions

During the melting of the Cordilleran ice sheet, two major volcanic eruptions appear to have taken place in the Glacier Peak area of the Cascade Range of NW USA. The Glacier Peak eruptions are of particular interest since they appear to have been highly explosive. For example, the eruptions resulted in the transport of significant quantities of ash at least 1000 km east of the volcano while pumice was deposited to a thickness of 2–3 m as far as 30 km downwind (Porter 1978; Beget 1982). In addition, approximately 3–5 km³ of pyroclastic and debris flows were produced as well as 5 km³ of airborne tephra (Beget 1982). The ages of the two eruptions are subject to some uncertainty, but they are generally considered to have taken place approximately between 12,750 and 11,250 BP (Mehringer *et al.* 1984; Porter 1978; Beget 1984). It is noteworthy that a significant Lateglacial ice advance, that culminated between 12,000 and 11,000 BP, took place in western North America soon after the Glacier Peak eruptions (see Porter 1978: 40). Beget (1983) noted that there was a correspondence between the time of the Glacier Peak eruptions and the global climate cooling of the Younger Dryas.

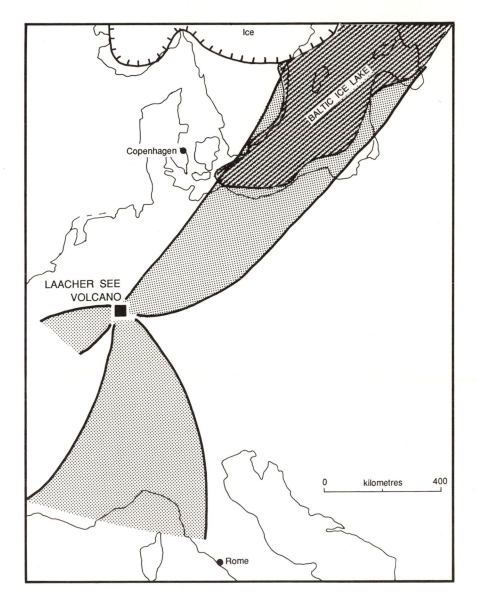

Figure 10.6 Distribution of Laacher See tephra. Large quantities of tephra may have been deposited in the Baltic Ice Lake (after Bogaard and Schmincke 1985).

195

INFLUENCE OF CLIMATE CHANGE ON
VOLCANIC ACTIVITY

The possibility that Late Quaternary environmental changes may have influenced the timing of volcanic eruptions is a relatively new one. Rampino *et al.* (1979) have argued that the redistribution of water that accompanied glaciation and deglaciation during this period may have given rise to both hydro–isostatic and glacio–isostatic readjustments. Crustal adjustments would have been most active along plate margins and major lineament–fault intersections. In this way, fault reactivation may have triggered subcritical magma bodies through the mechanism of magma mixing (Rampino *et al.* 1979). Hall (1982) has argued on several grounds that rapid deglaciation may frequently have led to the occurrence of both faulting and volcanic activity. Hall considers that the very rapid ice thinning during deglaciation could have led to considerable changes in the stress, strain, and strain rates in the bedrock and that fault lines generated by these processes may later have acted as conduits for ascending magma. Similarly, Sejrup *et al.* (1989) maintain that deglaciation may have been associated with an increased thermal gradient in the crust as well as increased pressures within individual magma chambers that eventually led, on occasions, to explosive volcanic eruptions.

Certainly, several very large volcanic eruptions took place in association with regional deglaciation. Thus, the explosive eruption of Mt Katla in southern Iceland at 10,600 BP took place from beneath a rapidly thinning ice mass. Similarly, the highly explosive eruptions of Glacier Peak between 12,750 and 11,250 BP in the North Cascade Range took place in association with the rapid thinning of Cordilleran ice. More recently, Sejrup *et al.* (1989) have shown that an extremely large eruption took place on Iceland during the last interglacial (substage 5e) and have argued that this was due to deglaciation of the Icelandic ice cap. Sejrup *et al.* (1989) point out that the eruptions during substage 5e as well as those associated with the 10,600 BP event were several orders of magnitude larger than those that took place during the remainder of the Holocene.

Sejrup *et al.* (1989) have also described a scenario whereby ice sheet build-up results in the downwarping of the upper crust and consequently a reduction in volcanic activity. In addition, Anderson (1987) has described how rapid changes in ocean water pressure may also influence volcanic activity. Thus he argues that a fall in relative sea level in a particular area may be accompanied by an increase in volcanism whereas a rise in sea level would result in a decrease in volcanic activity.

SUMMARY

In this chapter, the intention has been to show that the tectonically active regions of the world have been the focus of some cataclysmic volcanic eruptions during the last 125,000 years. It is not known if some of the stratospheric dust veils

196

caused by some of the largest eruptions may have induced global cooling sufficient to trigger or accelerate glaciation. It is possible, however, that certain
climatic regimes may have been more sensitive than at present to the injection of
large volumes of volcanic aerosols into the upper atmosphere. For example, the
Toba and Atitlan eruptions of 75,000 and 84,000 BP took place during periods

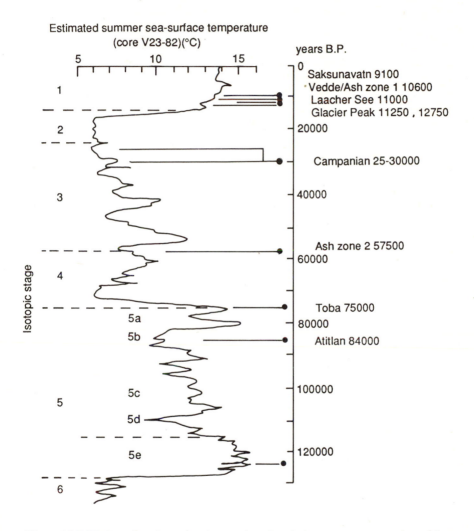

Figure 10.7 Timing of major volcanic eruptions in relation to a reconstruction of Late
Quaternary summer sea surface temperature changes for the North Atlantic from a core
sampled at circa 53°N 22°W. The oxygen isotope stages and substages are also shown
(after Sancetta *et al.* 1973). No correlation between eruptions and sea surface temperatures is implied.

of rapidly declining Milankovitch calorific insolation and increased global ice accumulation.

The Toba eruption in Sumatra appears to have been by far the most cataclysmic Late Quaternary eruption and it remains for future research to determine the influence of this event on global climatic cooling and ice sheet build-up during the oxygen isotope stage 5/4 transition. The great magnitude of the eruption when compared with other large eruptions as well as the immense volume of released magma leads one to the conclusion that the Toba eruption was probably one of the most catastrophic volcanic events to have occurred during the entire Quaternary. In the future, it may prove possible to estimate the climatic impact of such large eruptions through the simulation of large-scale ash injections within advanced general circulation models of the Earth's atmosphere.

It is clear that a number of high-magnitude volcanic eruptions also took place during the Lateglacial (Figure 10.7). Of these, the most explosive appear to have been the Glacier Peak eruptions and the Katla eruption in southern Iceland. At present the influence of these eruptions on Younger Dryas climatic cooling is not known although it is unlikely to have been large. In NW Europe, however, the ash layers deposited by the Katla and Laacher See eruptions have assumed a quite different significance for scientists. This is because the widespread occurrence of these ash layers has enabled the correlation of Lateglacial marine and continental sediment sequences across Europe and the North Atlantic. A promising area of future research will be to look for evidence of deposition of Vedde and Laacher See tephra within sediments of the Baltic Ice Lake (see Chapter 5). If this were possible, it would then be possible to link a European and North Atlantic tephrachronology with the Scandinavian varve chronology, thereby providing an excellent control on the dating of regional Lateglacial environmental changes.

11

CRUSTAL AND SUBCRUSTAL EFFECTS

INTRODUCTION

Consideration of the crustal response of the Earth to loading and unloading by ice and water masses has assisted greatly in our understanding of the structure of the Earth's interior, which in turn has significantly enhanced understanding of plate tectonics. A mobile lithosphere is a fundamental element in the thermal convective models used in theories of continental drift (Jarvis and Peltier 1982). Particular controversy has focused on whether or not there is a relatively low-viscosity asthenosphere that enables subcrustal compensation whenever loads are applied to or removed from the lithosphere (Table 11.1).[1]

During the Late Quaternary, the growth and decay of ice sheets and related changes in global sea level led to the redistribution of mass across the surface of the Earth. The rates at which the lithosphere and asthenosphere responded to these changes are not known. It is probable that many of these changes took place at a relatively slow rate. However, on some occasions (e.g. during the drainage of very large ice-dammed lakes), the changes may have been much more rapid. Rapid changes may also have been associated with earthquake activity and with faulting along pre-existing zones of geological weakness.

In areas of active plate collision, Late Quaternary crustal movements assumed a quite different significance. During this period, these areas have been sites of major earthquakes and volcanic eruptions while fault activity has also been widespread. Consequently, crustal deformation has played an important role in the evolution of individual landscapes.

MANTLE VISCOSITY AND CRUSTAL DEFORMATION

Several attempts have been made to use sea level change data and isostatic rebound patterns to evaluate the mechanical response of the Earth to surface loading and to constrain the melting histories of Late Quaternary ice sheets (Lambeck 1990; Lambeck *et al.* 1990). These data have also been used to provide information on the nature of the viscosity profile of the mantle. This is of great relevance since it has been suggested that mantle convective circulation

Table 11.1 Viscosity profile through the Earth's interior

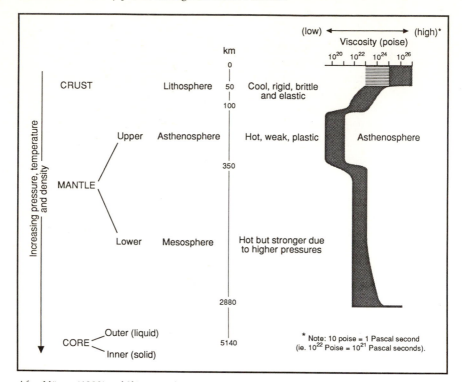

After Mörner (1980) and Skinner and Porter (1987). Not to scale.

cannot transfer material across the interface between the lower and upper mantle (Peltier 1981).

Peltier and Andrews (1983) have focused attention on two different mantle viscosity models. The first of these is based on the premise that the mantle has a relatively uniform viscosity with depth. Their second model is characterised by an increase in viscosity at the top of the lower mantle. Peltier and Andrews (1983) have argued that the empirical data on crustal uplift due to unloading of the Laurentide ice are best explained by a model of uniform mantle viscosity in which loading and unloading processes are compensated by global-scale mantle readjustments. A quite different view, based on data from Fennoscandia, has been suggested by Mörner (1980a) who considered that the Fennoscandian uplift data are compatible with a low-viscosity asthenosphere beneath the lithosphere. This implies that the processes of glacial loading and unloading are associated with elastic deformation of the lithosphere and plastic deformation within a compressed asthenosphere. In this way, loading and unloading of the lithosphere by ice sheets lead to compensatory deformation that takes place entirely within the ice sheet area and surrounding forebulge regions.

200

In most circumstances, the magnitude and dimensions of crustal deformation due to the imposition of an ice load are calculated theoretically once the dimensions of the ice sheet are known. This, in turn, is dependent on accurate geomorphological and stratigraphic observations as well as precise radiometric dating of individual ice marginal moraines and raised shorelines. Moreover, the occurrence of dated ice-marginal moraines and raised shorelines provides independent estimates of the dimensions of the former ice sheet. Geophysical models use such data to calculate a mantle viscosity profile that is compatible with the field data. At present, therefore, there is a great deal of uncertainty concerning the relationship between mantle viscosity and the forces that drive plate tectonics and mantle convection. In the future, the modelling of glacial rebound patterns and sea level changes associated with the melting of Late Quaternary ice sheets might eventually provide some answers to some of these questions (Lambeck 1990; Lambeck *et al.* 1990).

CRUSTAL DEFORMATION CAUSED BY ICE SHEETS

Ice sheet loading

The elastic deformation of the lithosphere that results from ice sheet growth is approximately proportional to the ratio between the density of the ice sheet and the underlying crust. This glacio-isostatic or crustal depression is greatest where the ice is thickest and declines towards and beyond the ice sheet margin. At greater distances beyond the ice margin the crust is thought to be subject to upwarping that results in the development of a forebulge (Walcott 1970) (Figure 11.1). This is due to the lateral transfer of subcrustal material from beneath the ice load. Conversely, ice sheet melting is associated with a return of this material towards the centre of glacio-isostatic uplift. At the same time, crustal depression takes place in the area of the forebulge.

It is likely that glacio-isostatic crustal depression associated with the development of individual Late Quaternary ice sheets was extremely complex. This is illustrated by the patterns of crustal depression presently associated with the Antarctic ice sheet (Drewry 1983) (Figure 11.2). Here, glacio-isostatic depression due to the present ice load is associated with the formation of several separate centres of isostatic downwarping. A peripheral crustal forebulge is assumed to occur around Antarctica but its extent and magnitude are not known.

The patterns of glacio-isostatic depression caused by the growth of the last Laurentide and Eurasian ice sheets can be estimated by comparing the present patterns and rates of isostatic uplift. The most accurate reconstructions of glacio-isostatic depression, however, are gained from studies of isostatically uplifted and regionally tilted shorelines (Chapter 12). These show clearly, for example, that the Laurentide ice sheet during the last glacial maximum had multiple ice domes and that each was associated with a separate centre of crustal depression

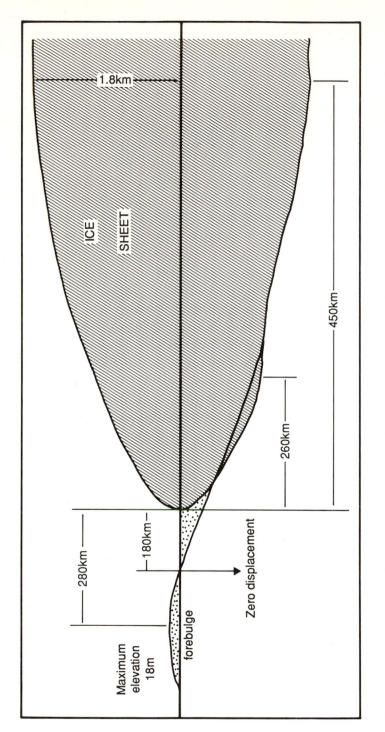

Figure 11.1 Idealised model of crustal deformation caused by an ice sheet. Note area of peripheral crustal forebulge (after Walcott 1970).

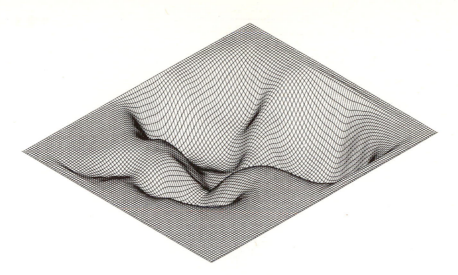

Figure 11.2 Isometric view showing complex nature of glacio-isostatic depression of Antarctica. The figure is oriented with the Antarctic Peninsula at the bottom (after Drewry 1983).

(Andrews 1970). By contrast, the pattern of crustal depression over Fenno-scandia was much simpler with a single centre of isostatic downwarping (Mörner 1980a).

Ice sheet unloading

In areas of glacio-isostatic uplift, the rate of crustal rebound varies according to the thickness of the former ice cover. The greatest rates of uplift occur in areas where the ice was formerly thickest. In those areas where uplift outpaces glacio-eustatic sea level rise, raised shorelines are produced. Since the rate of uplift diminishes with increasing distance from the uplift centre, individual synchronous raised shorelines exhibit a regional tilt and decline in altitude at greater distances away from the uplift centre. Radiometric dating of these shorelines coupled with detailed measurements of their altitude variations and distribution can therefore be used to develop numerical models of glacio-isostatic crustal uplift and subcrustal flow. Equally they provide valuable information on the dimensions of former ice sheets and their melting histories (Eronen 1983).

The initial thinning of an ice sheet is accompanied by a period of restrained rebound beneath the ice cover (Andrews 1970) (Figure 11.3). The nature of crustal rebound during this period is difficult to determine except by the use of theoretical models. The calculation of restrained rebound is problematic since it assumes that the crust was in isostatic equilibrium prior to the build-up of the last

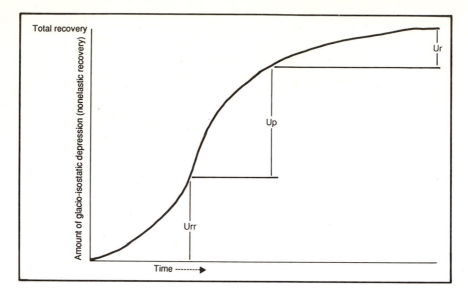

Figure 11.3 The three periods of glacio-isostatic recovery: 1. restrained rebound (Urr), 2. postglacial uplift (Up) and 3. residual uplift (Ur). The form of the curve is schematic (after Andrews 1970).

ice sheet. Indeed, it is likely that areas of the crust were still undergoing uplift from a previous period of deglaciation when the build-up of the last ice sheets took place. The second phase takes place once individual areas become deglaciated. This period of postglacial uplift is accompanied by the formation of raised shorelines.

The patterns and processes of postglacial uplift may become greatly complicated by the readvance and thickening of ice sheets during deglaciation. For example, the renewed build-up of ice in Scandinavia during the Younger Dryas (Chapter 5) is not only thought to have retarded crustal uplift but may even have caused crustal redepression (Anundsen 1985). Finally, there is a component of residual uplift that has not yet taken place and which may be considerable. For example, in southern Hudson Bay, Canada, between 120 and 160 m of uplift remains to be completed while in more peripheral areas (e.g. Nova Scotia) the value may be as high as 40 m.

In Scandinavia, the patterns of present uplift display a concentric zonation and correlate approximately with the distribution of free air gravity anomalies (Figure 11.4).[2] Similarly in North America, there is an approximate correspondence between the former distribution of ice sheets and the pattern of free air gravity anomalies. In general, the distribution of free air gravity anomalies coincides with areas where glacio-isostatic uplift is incomplete although this relationship is additionally complicated due to the effects of rock structure and topography (Wu and Peltier 1983; Vita-Finzi 1986).

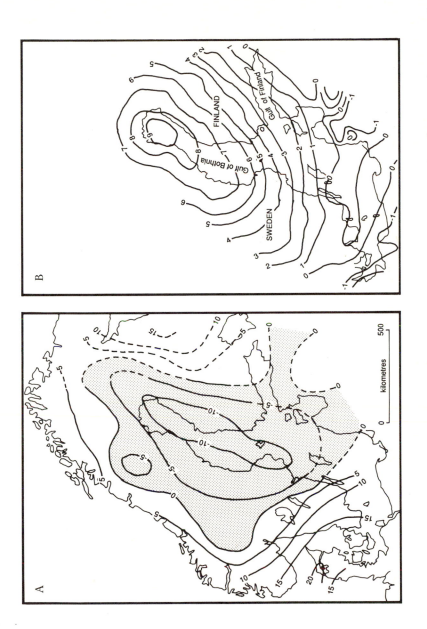

Figure 11.4 A. Regional gravity anomaly map (mgal) for Fennoscandia (after Balling 1980). B. Present-day relative crustal movements (mm yr^{-1}) in the Baltic area (after Eronen 1983). Note the similarities between the two diagrams.

Deformation of marine shorelines due to ice unloading

The changing nature of postglacial glacio-isostatic uplift can be determined if it is possible to identify the nature of the regional eustatic sea level curve for any particular area. In this way, curves of relative sea level change derived from areas affected by glacio-isostatic uplift can be compared with the regional eustatic sea level curves in order to isolate the component of isostatic uplift (e.g. Andrews 1970; Mörner 1980a,b). In general the calculated isostatic displacements are characterised by rates of uplift that exhibit an initial increase during the period of restrained rebound and thereafter decrease exponentially during and following ice sheet deglaciation.

Individual marine shorelines formed during ice sheet deglaciation decline in altitude away from the centre of isostatic rebound (Figure 11.5). In general the shorelines that exhibit the greatest regional tilt are the oldest although this pattern may be complicated considerably by the occurrence of renewed ice build-up and isostatic redepression that interrupted the overall pattern of ice thinning

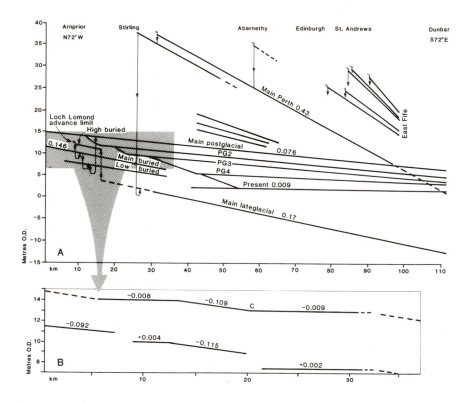

Figure 11.5 Shoreline height–distance diagram for the western Forth valley, eastern Scotland. Inset shows nature of localised shoreline dislocation. Regional shoreline gradients are given in m/km (based on Sissons 1972, 1974a).

Plate 14 Raised shoreline staircases of early Holocene age, Great Bear Lake, North West Territories. Photo A.K. Lobeck. This aerial photograph © 1958 Her Majesty the Queen in Right of Canada, reproduced from the collection of National Air Photo Library with permission of Energy, Mines and Resources Canada.

and retreat. The calculation of the pattern of glacio-isostatic recovery can be achieved for individual areas by subtracting the curve of relative sea level change (based on raised shoreline data) from a regional eustatic sea level curve (Mörner 1980a,b). In general the curve of postglacial uplift exhibits a pattern of decreasing isostatic uplift. However, on rare occasions the pattern of uplift is interrupted by fault activity (Gray 1974; Sissons 1983) (Figure 11.5).

Deformation of glacial lake shorelines due to ice unloading

During ice sheet deglaciation, differential glacio-isostatic uplift led to the regional tilting of shorelines that had previously formed along the margins of ice-dammed lakes. The clearest and most pronounced patterns of shoreline deformation are evident for the largest lakes. Of these, particularly well-documented examples are known for the large ice-dammed lakes that bordered the southern margins of the Laurentide and Scandinavian ice sheets. For example, very detailed information is available for glacial Lake Agassiz where as many as 30 tilted shorelines provide a record of changes in uplift and lake level, due to fluctuations in the former ice

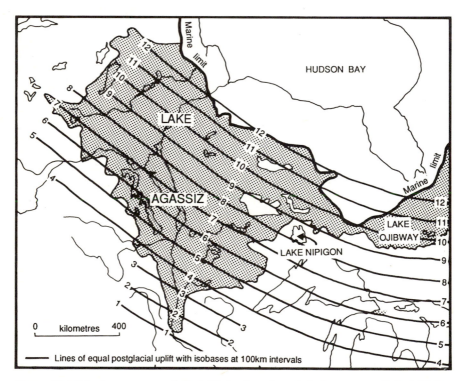

Figure 11.6 Maximum extent of glacial Lake Agassiz and the late stages of glacial Lake Ojibway (Vincent and Hardy 1979) showing the pattern of glacio-isostatic shoreline deformation (based on Teller and Thorleifson 1983).

margin and the use of different overflow routes (Teller and Thorleifson 1983). Crustal rebound has deformed the surface considerably since deglaciation, with a maximum rise in shoreline elevation towards the north-northeast (Figure 11.6).

Faulting and earthquake activity related to deglaciation

There is a growing body of evidence to support the view that the highly dynamic nature of the lithosphere during rapid glacio-isostatic uplift causes crustal instability. Indeed, Mörner (1980a) has argued that large earthquakes and faulting seem to have been characteristic of the last period of ice sheet deglaciation in Fennoscandia. Mörner (1980a) cited early work by De Geer who provided evidence for the occurrence of large earthquakes in the Stockholm area during deglaciation. Numerous fault scarps related to glacio-isostatic uplift have also been described from Fennoscandia (Mörner 1980a). Indeed, one of these in northern Sweden is 150 km long and up to 10 m high (Lundqvist and Lagerback 1976). In Scotland, Ringrose (1989) has provided evidence for widespread palaeoseismicity and earthquake activity during deglaciation.

It is not surprising that high seismic activity took place during deglaciation since the lithosphere is inhomogeneous and fractured in many places by deep fault lines. In addition, the processes of glacio-isostatic crustal deformation (both loading and unloading) are likely to have greatly increased both crustal stress and strain rates (Mörner 1980a,b). Thus whereas rapid uplift may have triggered fault activity along lines of pre-existing geological weakness, there may also have been fault activity that took place as a direct consequence of rapid glacio-isostatic uplift. The passive and normally stable continental plates upon which the mid latitude ice sheets developed appear, therefore, to have been characterised by tectonic instability during ice sheet melting (Bjorck and Digerfeldt 1986).

Crustal deformation due to water loads

Crustal deformation of the lithosphere due to water loading and unloading may take place as a result of sea level changes but it may also take place as a result of changes in lake levels. Given that tidal changes may trigger seismic activity (Melchior 1978), it is not surprising that global sea level changes caused by the melting of ice sheets caused substantial hydro–isostatic crustal deformation (Hopley 1983). To date, the hydroisostatic deformation of the ocean floor due to sea level changes has largely been established on a theoretical basis although empirical data do exist (e.g. Hopley 1983). Thus, the calculations of global relative sea level lowering for the last glacial maximum incorporate a fixed component of hydroisostatic deformation (Mayewski *et al.* 1980) (Table 4.3). In this way, low glacio-eustatic sea levels during the last glacial maximum were associated with a compensatory oceanic crustal uplift. Conversely, the rise in sea level that has since taken place has been associated with crustal subsidence.

The hydroisostatic influence of lakes on crustal dynamics is demonstrated clearly by the shorelines of 'pluvial' Lake Bonneville (Gilbert 1890; Crittenden

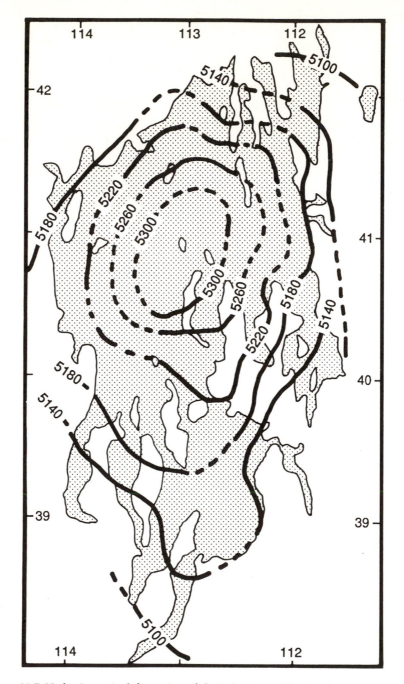

Figure 11.7 Hydro-isostatic deformation of the Lake Bonneville shoreline (in feet) (after Crittenden 1963).

210

1963) (Figure 11.7) (see also Chapter 7). During its most extensive phase, when the lake had a maximum depth of 355 m, a shoreline was produced at 1,570 m above sea level. Differential tilting has since led to the occurrence of this shoreline up to a maximum altitude of at least 50 m above its original level (Figure 11.7). The Bonneville shoreline data are significant since they indicate that other large lakes formed during the Late Quaternary may have exerted similar hydroisostatic deformation upon the underlying crust.

A characteristic of the ice sheets that were produced during the last glaciation is that they were frequently bordered at some time during their history by very large glacier-dammed lakes. During deglaciation many lakes were subject to catastrophic drainage due to the breaching of ice barriers by lake waters (see Chapter 5). Surprisingly, some of the clearest evidence for crustal movement associated with the catastrophic drainage of ice-dammed lakes is available from one of the smallest – Glen Roy in the Scottish Highlands (Plate 9).

During the Younger Dryas stadial, a 5 km^3 volume of lake water was impounded by glacier ice. The eventual drainage of the water occurred subglacially as a catastrophic flood and was associated with widespread landslide activity and faulting (Sissons and Cornish 1982). Ringrose (1989) considered that these events, in turn, were related to earthquake activity, perhaps caused by sudden lake drainage. Sissons and Cornish (1982) argued that if such palaeo-seismic activity was related to the drainage of such a comparatively small lake, similar processes may have taken place on a much greater scale in the areas where large ice-dammed lakes were subject to rapid drainage.

Crustal deformation due to sediment loading

The greatest rates of isostatic depression caused by sediment loading occur in deltaic regions. For example, in the Mississippi delta area, sediments deposited approximately 10,000 years ago occur at −200 m when regional eustatic sea level was at −35 m. Fairbridge (1983) has explained this anomaly of 165 m as partly the result of sedimentary compaction (15 m) but mostly due to tectonic subsidence (150 m) that has taken place at an average rate of 15 mm per year since the beginning of the Holocene.

CRUSTAL DISTURBANCE ALONG ACTIVE PLATE MARGINS

The uplift of mountain masses along active plate margins has led to some quite spectacular changes in the Earth's landscape during the Late Quaternary. In areas where large-scale crustal uplift has taken place, significant changes in regional climate may have resulted. Perhaps the most important of these is the Tibetan Plateau where as much as 3,000 m of uplift may have taken place during the Quaternary with much of this uplift having occurred during the Late Quaternary (Shi *et al.* 1986; Tungsheng *et al.* 1986). It is possible that these changes may

Plate 15 Fault-displaced palaeosols (darker areas) developed on airfall pumice layers near Guatemala City. Considerable tectonic activity occurs in this area as a result of collision between the Cocos and Caribbean plates. Photo S.C. Porter courtesy John Wiley and Sons.

have resulted in a progressive reduction in snow precipitation, the weakening of monsoonal circulation and the development of progressively smaller Tibetan glaciers.

Considerable Late Quaternary uplift has also taken place in parts of the southern hemisphere. For example, Clapperton (1987) has observed that the Quito area of Ecuador may have been uplifted by approximately 150 m during the last 50,000 years. In South Island, New Zealand, convergence of the Pacific and Australian plates since the last interglacial has resulted in uplift rates of between 3.2 and 7.8 m per thousand years (Bull and Cooper 1986) (e.g. Plate 15).

The scale of Late Quaternary crustal deformation in tectonically active areas can also be determined from studies of raised shorelines (see also Chapter 12).

For example, the dated coral reef 'staircase' sequence of elevated beaches on the Huon Peninsula in Papua and New Guinea has been used to show that average uplift in this area since the last interglacial has been 2 mm per year (Chappell 1974a). Uplift between the last interglacial and 80,000 BP appears to have been more rapid than it was during the last 80,000 years. However, the style of deformation has remained essentially unaltered throughout the Late Quaternary despite the proximity of the area to the boundary of the West Pacific and Australian plates (Vita-Finzi 1986). Chappell (1983, 1987) provided additional evidence from this area and suggested that the uplift, rather than being gradual, has been episodic and probably caused by individual seismic events.

Estimates of Late Quaternary uplift and subsidence can often be made through the altitude measurement of shorelines formed during the culmination of the last interglacial (substage 5e). In tectonically stable areas the shoreline usually occurs between 2 and 8 m. Hence, its occurrence at altitudes above or below this range implies disturbance by tectonic activity. Pronounced crustal movements are indicated for parts of the Mediterranean where marine deposits of last interglacial age occur at +120 m in Calabria, Sicily, and at −70 m near Venice (Pirazzoli 1987). Similarly marine deposits in southern Greece of last interglacial age show a progressive decrease in altitude from the southernmost part of the Hellenic arc (where they occur up to +100 m) towards the north where they occur slightly below sea level in the Aegean (Pirazzoli 1987). Similarly, in Japan the altitude of the last interglacial shoreline ranges between +10 m and +200 m above sea level, the variation in altitude being due to the long-term subduction of the Philippine Sea plate beneath the Eurasian plate (Ota and Machida 1987). In some areas of central Honshu along the Sea of Japan, the tilting of small blocks of shoreline of last interglacial age (isotope substage 5e) implies an average Late Quaternary uplift rate of up to 1.5 m per thousand years.

SUMMARY

The spectacular environmental changes that took place during the Late Quaternary should not be considered as events whose influence was confined to the Earth's surface. The major changes in the redistribution of mass known to have taken place (e.g. ice sheet growth and decay, sea level change, lake evolution) exerted an important influence on the underlying crust and mantle at different scales. At a continental scale, the development of the Laurentide and Eurasian ice sheets exerted elastic deformation of large areas of passive continental plate. On a smaller scale, significant hydroisostatic deformation of the lithosphere is illustrated by the development of the Lake Bonneville shorelines. Similarly, the drainage of as little as 5 km³ of water from the ice-dammed lake in Glen Roy, Scotland, resulted in widespread landslide activity and faulting. These examples demonstrate the sensitivity of the crust to the imposition and release of loads.

Numerical modelling of former ice sheet histories and sea level changes has led to the development of models of mantle convection that have supplemented existing geophysical models. These have highlighted a controversy between two different models of mantle convection. The occurrence of a mantle of uniform viscosity beneath the (high-viscosity) lithosphere (e.g. Peltier) implies global readjustment to isostatic loads. In contrast, the occurrence of a low-viscosity asthenosphere (Mörner) in the upper mantle implies a more regional adjustment to ice sheet loading and unloading.

Numerous numerical models have been developed that seek to reconstruct patterns of ice sheet glacio-isostatic deformation that are compatible with the field evidence of former shoreline changes and ice limits. Consideration of the inferred glacio-isostatic deformation caused by the present Antarctic ice sheet suggests that we should be cautious in our interpretations since the response of the lithosphere to ice sheet loading is highly complex. Similarly, lithosphere rebound caused by the melting of the last major ice sheets may have led to wide-spread seismic activity – much of which remains unknown. Elsewhere, in areas of active plate movement, seismic activity has been commonplace throughout the Late Quaternary. In some areas, evidence of warped shorelines demonstrates considerable uplift (up to 200 m) since the last interglacial. The Tibetan Plateau appears to have been subject to crustal uplift on a grand scale. Here, uplift may have been sufficiently great to have altered patterns of regional atmospheric circulation.

NOTES

1   The lithosphere consists of the outer 100 km of the solid Earth where rock is harder and more rigid than in the underlying asthenosphere. The asthenosphere is the region of the upper mantle, occurring at depths between 100 and 350 km below the Earth's surface, where rock becomes plastic and is more easily deformed. In the lower mantle (or mesosphere) the higher pressure, temperature and density result in the occurrence of rock that is much less easily deformed than in the overlying asthenosphere (Table 11.1).

2   These are regional variations in the force of gravity that have been adjusted for variations in elevation (see Skinner and Porter 1987).

12

LATE QUATERNARY SEA LEVEL CHANGES

INTRODUCTION

The dynamic nature of Late Quaternary sea level changes was largely the result of several interacting isostatic, eustatic and tectonic processes. Isostatic changes relate to the condition of equilibrium established by the Earth's lithosphere, which essentially 'floats' on the asthenosphere (Fairbridge 1983). Eustatic sea level changes represent changes in the shape and level of the surface of the world's oceans that exist in equilibrium with the gravity field of the Earth. Vertical changes in this surface may be due to water volume changes (glacio-eustatic), to changes in the shape of the equipotential surface (geoidal-eustatic) and to changes in the shape of the ocean basins (tectono-eustatic) (Fairbridge 1983). Tectonic changes, by themselves, refer to vertical crustal movements that may locally affect the position of relative sea level (Table 12.1).

The major fluctuations in global sea level during the Late Quaternary were glacio-eustatic in origin and were the result of changes in global ice volume. However, tectono-eustatic changes, caused by changes in the shape of ocean basins, may also have been responsible for some of the observed sea level variations. There may also have been important variations in the geoidal-eustatic sea surface during the Late Quaternary although these are not known. Some of these changes were due to regional variations in the Earth's gravity field that caused gravitational attraction of ocean water to adjacent ice sheets and thus raised relative water levels in these areas. Glacio-isostatic and hydro-isostatic effects caused by the alternate loading and unloading of Late Quaternary ice sheets on the lithosphere also led to complex regional fluctuations in relative sea level (Figure 12.1).

Despite our knowledge of the eustatic and isostatic processes responsible for Late Quaternary sea level changes, it has proved astonishingly difficult to reconstruct the pattern of sea level changes that actually took place. In several areas of the world where there is empirical field evidence for former sea level changes, attempts have been made to reconstruct past sea level variations. For example, Bloom (1967) proposed that fossil shorelines that occur on guyots and pinnacle islands in areas of oceanic crust might provide an accurate record of Late Quaternary global eustatic sea level changes. He argued that such areas would be unaf-

Table 12.1 Processes controlling the factors giving rise to the three different types of eustasy

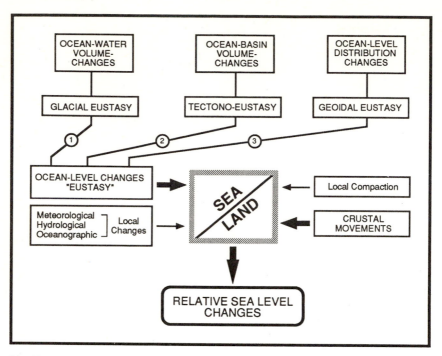

After Mörner (1980).

fected by hydro-isostatic loading and unloading, and that the islands would respond to water level changes in a similar manner to a dipstick. However, Mörner (1987) has subsequently shown that such areas may have been affected by substantial geoidal sea surface changes during the Late Quaternary and hence are of limited value in the reconstruction of former sea levels. The most detailed record of past sea level changes, however, has been derived from the oxygen isotope record of benthonic foraminifera (Chapter 3). Many authors have argued that since the oxygen isotope record obtained from ocean sediments is primarily a curve of global palaeoglaciation, it is also a curve of glacio-eustatic sea level changes (see p. 10).

The radiometric dating of uplifted and submerged coral reefs, known to have formed within particular ranges of water depth, has also been used to reconstruct Late Quaternary eustatic sea level fluctuations. For example, Chappell (1974a) and Bloom *et al.* (1974) have used data from flights of uplifted coral reefs in New Guinea to produce a Late Quaternary eustatic sea level curve. Similarly, Fairbanks (1989) has prepared a detailed eustatic sea level curve for the last 18,000 years for Barbados using data on submerged coral of known ages (see p. 233). In both of these examples, the calculation of the eustatic sea level curve is depen-

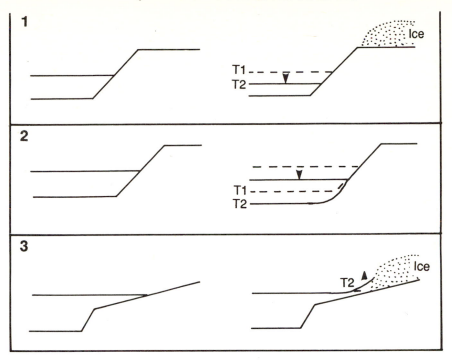

Figure 12.1 Three models to explain absolute, eustatic changes in sea level. 1. Glacial eustasy (changes in water volume). 2. Tectono-eustasy (changes in basin volume). 3. Gravitational mass attraction of water (changes in water distribution) due to the build-up and decay of ice sheets (after Mörner 1987).

dent on assumptions being made about the rates of vertical tectonic movements that have taken place in these areas during the respective time periods.

The reconstruction of former sea levels for individual areas is complicated by a number of processes the most important of which, apart from dating, are (1) glacial isostasy, (2) hydro-isostasy, (3) tectono-eustasy, (4) geoidal eustasy and (5) vertical crustal movements. These are considered in more detail in the following sections. In turn, this discussion is succeeded by a description of the nature of the sea level changes that are generally believed to have taken place during the Late Quaternary.

FACTORS AFFECTING LONG-TERM SEA LEVEL CHANGES

Glacial isostasy

In formerly glaciated areas, fossil shorelines are regionally tilted owing to differential glacio-isostatic uplift of the lithosphere following ice sheet decay (Plate

14). Since the amount and rate of isostatic recovery decreases in directions away from the former centre of uplift while the absolute change in sea level remains regionally uniform, the pattern of relative sea level changes varies greatly since it reflects glacio-isostatic movements of the crust as well as eustatic changes.[1] Many investigations of former relative sea level changes in areas affected by glacio-isostatic uplift have been used to reconstruct patterns of isostatic deformation caused by the growth and decay of ice sheets. These, in turn, have been used to identify past eustatic changes in sea level by separating the measured relative sea level change from the component attributable to vertical glacio-isostatic deformation of the crust (Mörner 1971).

In theory, the patterns of relative sea level change known to have taken place in areas affected by glacio-isostatic deformation provide extremely valuable information on the response both of the Earth's crust and of global sea level to the growth and decay of Late Quaternary ice sheets. In practice, however, it is difficult to distinguish the components of relative sea level change that are attributable to glacio-isostatic deformation and glacio-eustatic fluctuation. This is because the formation of individual raised shorelines at particular altitudes also reflects the influence of other eustatic and isostatic factors. For example, the sea surface rise caused by the geoidal attraction of ocean water to ice sheets is usually not known although it may have been considerable. Similarly, it is almost impossible to determine the amount of crustal uplift directly attributable to the melting of the last ice sheets since crustal uplift may not have been completed prior to the last period of ice sheet growth. This is particularly important since the growth and decay of large ice sheets may have taken place on several occasions during the Late Quaternary. Thus, it may be appropriate to envisage the lithosphere and asthenosphere as having been subject to numerous highly dynamic glacio-eustatic and glacio-isostatic fluctuations throughout the Late Quaternary.

Hydro-isostasy

The hydroisostatic loading and unloading of water upon ocean floors and continental shelves, as a result of sea level changes, is poorly understood (Walcott 1972; Chappell 1974b). In theory, a long-term rise in sea level (e.g. during the last period of ice sheet deglaciation) is associated with a compensatory depression of crustal areas beneath the water bodies. Clearly, the tectonic setting of individual regions and the flexural rigidity of the lithosphere are important aspects to consider in this respect, especially for areas of continental shelf where such deformation might be expected to have occurred.

The possible magnitude of these changes is well illustrated by Chappell (1974b) who has argued that sea level rise during the last 7,000 years has resulted in an overall 8 m depression of the ocean floors and an average uplift of the continents of 16 m. By contrast, Hopley (1983) has shown for north Queensland, Australia, that the values for emergence and subsidence for this time period may have been considerably smaller, with the release of hydroisostatic stresses

Plate 16 Raised rock platform in quartzite, northern Islay, Scottish Hebrides. The seaward section of the platform is mantled by raised beach sediments deposited during regional ice sheet deglaciation. Photo D. Munro.

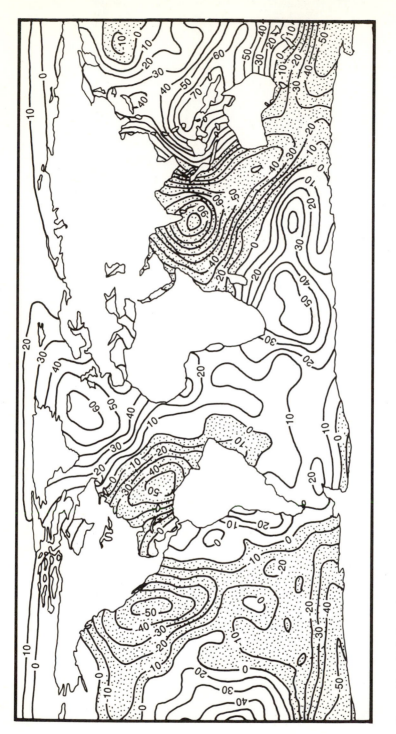

Figure 12.2 Global geoid map in metres with respect to the best-fitting ellipsoid (after Van de Plassche 1982). Negative areas are shown shaded.

having taken place along pre-existing fault lines. However, Mörner (1987) has challenged some of the key assumptions regarding hydro-isostatic deformation and has argued that whereas hydro-isostatic loading during eustatic sea level rise may induce subsidence of the ocean floor, it does not follow that hydro-isostatic unloading due to sea level fall would cause uplift of the ocean floor.

Tectono-eustasy

Bloom (1971) has argued that, due to sea floor spreading, ocean basins have been widening at an average rate of 16 cm per year during the Late Quaternary. He therefore considered that the world's oceans have accommodated an additional 6 per cent water volume since the last interglacial and hence shorelines of the present interglacial must be several metres lower than during the previous inter-glacial. The problem is complex since it can be argued that during the last 125,000 years, the total amount of sea-floor spreading that has occurred globally has been counterbalanced by reductions in ocean volume in areas of plate collision.

Geoidal eustasy

The present surface of the world's oceans is uneven due to regional variations in the Earth's gravity field (Gaposchkin 1973) (Figure 12.2). Consequently, the ocean surfaces of the Earth represent an equipotential surface of the Earth's gravity field, known as the geoid. There are considerable differences in geoidal sea surface altitudes; for example, satellite measurements have shown that there is a 180 m difference in sea level between the low level of the Maldive Islands (Indian Ocean) and the high level of New Guinea (Mörner 1976). Mörner (1980b) considered that there is a correlation between former geoidal changes and geomagnetic field anomalies for the Late Quaternary. He argued that some of these fluctuations may have been attributable to the redistribution of mass in the upper mantle and at the core/mantle boundary caused by the growth and decay of ice sheets. Clearly, therefore, there may be many (mostly unknown) factors responsible for geoidal surface changes. It is important to realise, however, that geoidal changes cause real but differential ocean level changes as opposed to changes in ocean basin (tectono-eustatic) or ocean water (glacio-eustatic) volume, which are worldwide in their effects (Van de Plassche 1982).

The most important changes in the Earth's gravity field that took place during the Late Quaternary were those caused by the gravitational attraction of ocean water to ice sheets (Figure 12.1). Consequently, the growth and decay of major ice sheets during the Late Quaternary may be considered as having caused repeated geoidal deformation of ocean surfaces. For example, Fjeldskaar and Kanestrom (1980) showed that the melting of the Younger Dryas ice sheet in Fennoscandia was associated with geoidal deformation of the sea surface. They argued that approximately 5 per cent of regional tilting of the Younger Dryas

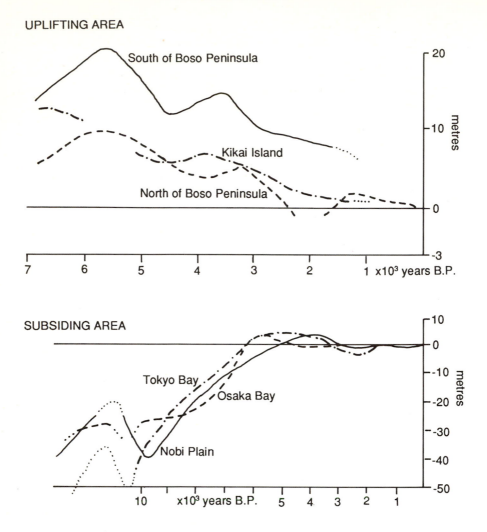

UPLIFTING AREA

South of Boso Peninsula

Kikai Island

North of Boso Peninsula

20

10 metres

0

-3

7   6   5   4   3   2   1 x10³ years B.P.

SUBSIDING AREA

Tokyo Bay

Osaka Bay

Nobi Plain

10

0

-10 metres

-20

-30

-40

-50

10   x10³ years B.P.   5   4   3   2   1

Figure 12.3 Holocene sea level curves for Japan showing effects of tectonic activity. In areas where tectonic uplift has taken place, there has been a relative marine regression. In subsiding areas, there has been a relative marine transgression (after Yoshikawa, Kaizuka and Ota 1981).

shoreline near Bergen, SW Norway, was attributable to such changes. During periods of ice sheet melting, ocean areas located considerable distances from the ice margin are likely to have undergone a geoidally induced rise in sea level.

Although the nature of the present geoidal surface can be determined from satellite measurements, it is extremely difficult to determine the nature of Late Quaternary variations in the geoid. Mörner (1976) has described possible patterns of geoid change that may have occurred during the Late Quaternary. His results suggest that the changes may have been extremely complex and consequently many published curves of sea level change may only be of local (rather than global) significance. These interpretations have cast doubt on the old view of eustasy that long-term sea level changes took place simultaneously on a worldwide basis. In summary, there is no such thing as a global eustatic curve of sea level change (Clark *et al.* 1978; Clark 1980; Mörner 1987).

Vertical crustal movements

The pattern of relative sea level changes is complex for most tectonically active island arc areas and areas of active plate collision. For example, Pirazzoli (1987) has shown that the Quaternary emerged shorelines in the Mediterranean have been subject to considerable vertical tectonic movements and hence are unreliable indicators of former regional sea level change (Hey 1978). Similarly, the raised shore features in Japan (Yonekura and Ota 1986; Ota and Machida 1987) and the Aleutian Islands (Black 1980) illustrate clearly the importance of major tectonic activity during the Late Quaternary (Ota 1987) (Figure 12.3).

However, tectonic uplift is not confined to active plate margins. For example, the occurrence of Late Quaternary raised beaches in the Falkland Islands (a passive margin of the South American continental plate) at altitudes up to +60 m has been attributed to progressive tectonic uplift (Clapperton and Roberts 1986). These authors have argued on several grounds that the Falkland Islands may have been tectonically uplifted at an average rate of 1 m per 6,000 years during the last 120,000 years. Similarly, Ota (1986), in a study of Late Quaternary tectonic deformation of raised marine terraces on shorelines bordering the Pacific, has shown that most areas characterised by plate convergence have been subject to considerable uplift, typically 2–3 m per thousand years throughout the last 125,000 years.

Late Quaternary subsidence is also locally important in influencing the record of former sea level changes. Considerable subsidence is especially evident in major delta environments and sea level curves derived from such areas clearly reflect these processes. This is well illustrated for Venice where shoreline sediments of last interglacial age occur at altitudes as low as −70 m (Pirazzoli 1987) (see also Chapter 11).

LATE QUATERNARY RELATIVE SEA LEVEL FLUCTUATIONS

Sea level changes between 125,000 and 18,000 BP

The most recent estimate of glacio-eustatic sea level fluctuations has been provided by Shackleton (1987) based on oxygen isotope studies of benthonic and

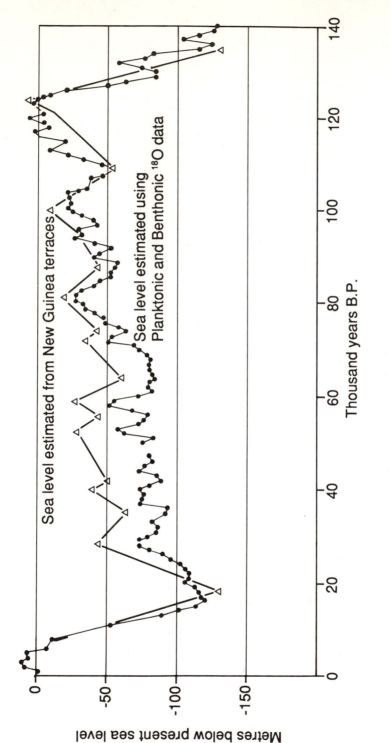

Figure 12.4 Oxygen isotope global sea level history compared with sea level data from New Guinea (after Shackleton 1987).

planktonic foraminifera (Figure 12.4). Shackleton's curve is important since it is calculated independently of emerged or submerged shoreline data. The results of several empirical studies of Late Quaternary sea level changes show remarkable correspondence to Shackleton's sea level curve. For example, the inferred pattern of Late Quaternary sea level changes for Majorca depicts a stepwise lowering of sea level since the last interglacial (when sea level was higher than present) until after the last glacial maximum when it began to rise (Butzer and Cuerda 1962). Similar curves of Late Quaternary sea level fluctuations have also been determined for Barbados (Mesolella *et al.* 1969), New Guinea (Bloom *et al.* 1974; Chappell 1974a), Haiti (Dodge *et al.* 1983), Japan (Ota and Machida 1987) and for Beringia, Alaska (Hopkins 1973).

Of particular interest is the tectonically active Huon Peninsula, New Guinea, where a series of spectacularly developed emerged coral reef terraces occur. Bloom *et al.* (1974) and Chappell (1974a) dated the various reef terraces and attempted to produce a eustatic sea level curve for the area by using an additional curve of 'uplift correction' (Figure 12.4). The sea level curve for New Guinea is particularly remarkable in the degree of correspondence between it and the curve of Shackleton (1987) (Figure 12.4). The New Guinea data (as well as the Barbados and Haiti data) suggest that during isotope substages 5a and 5c eustatic sea level was in the order of 20 m lower than it was during the peak of the last interglacial (substage 5e). More significantly, the oxygen isotope record when compared with the New Guinea data implies consistently lower sea levels between oxygen isotope substage 5c and stage 2 (Shackleton 1987).

The New Guinea, Barbados and Majorca relative sea level curves demonstrate that the accuracy of the reconstructed sea level curves is dependent on information on the tectonic history of the region. In the case of New Guinea, the reliability of the sea level curve is dependent upon a uniform pattern of Late Quaternary land uplift having taken place. That this is often not the case is well illustrated for Japan where Ota and Machida (1987) have shown that the raised beaches have been affected by non-uniform uplift during the last 100,000 years.

One of the remarkable aspects of the Barbados, Haiti, New Guinea and Japan sea level data is that there appears to be some correspondence between the periods of high sea level at 125,000 BP, 105,000 BP and 80,000 BP and the periods of maximum Milankovitch insolation for the Late Quaternary suggesting some cause–effect relationship (Broecker *et al.* 1968). Thus it may be argued that these periods of high sea level, since they are primarily glacio-eustatic in origin, are primarily driven by global ice sheet fluctuations that, in turn, are due to Milankovitch forcing (see Chapter 13).

Closer inspection of the Shackleton (1987) Late Quaternary sea level curve poses some fascinating additional questions when the curve is compared with the inferred worldwide glaciation chronology. The first aspect concerns environmental changes during and following the culmination of the last interglacial (isotope substage 5e). One possibility is that the West Antarctic ice sheet may have disintegrated during the isotope substage 5e peak of interglacial warmth, and that these changes caused a global sea level rise reaching up to 6 m above present (Stuiver *et al.* 1981). By contrast, Mercer (1968) considered that disintegration of the West Antarctic ice sheet may have taken place as a result of the higher sea levels and higher air temperatures of this period. Under such circumstances, one may speculate that the major period of global ice build-up at 115,000 BP (see Chapter 4) may have been related in some way to the climatic changes indirectly caused by ice sheet disintegration and the influx of glacial meltwaters into the world's oceans.

Another important issue concerns the sea level response to the melting of the ice sheets that had accumulated during the transition between oxygen isotope stages 5 and 4. Lundqvist (1986b) has argued that much of Scandinavia may have been deglaciated by 60,000 BP while Fulton (1986) has observed that the Cordilleran ice sheet may also have disintegrated during this period. Similarly, Andrews *et al.* (1986) concluded that the Laurentide ice sheet may also have been substantially reduced in volume at this time. The implication, therefore, is that such continental-scale ice melting should have resulted in a substantial glacio-eustatic rise in sea level. Support for this view is provided by Shackleton (1987), whose data suggest that a global sea level rise of approximately 30–40 m appears to have occurred between 60,000 and 55,000 BP.

Glacio-isostatic influences prior to the last glacial maximum

The pattern of relative sea level changes since the last interglacial is particularly complex for those areas overwhelmed by and peripheral to major ice sheets. In theory, the loading of areas by ice sheets results in deformation of the Earth's crust not only in those areas beneath the ice sheets but also in those areas that border them (see Chapter 11).

For the most part, the patterns of glacio-isostatic deformation associated with the growth and decay of the major northern hemisphere ice sheets during oxygen isotope stages 5, 4 and 3 are not known. Fragmentary information is available for Western Norway where marine sediments at 200–220 m deposited during and after the Skjonghelleren Stadial (see Chapter 4) point to a major glacio-isostatic depression associated with ice sheet build-up. In addition, Andrews *et al.* (1986) have shown that the occurrence in parts of Arctic Canada of high-level marine deposits, considered to have been deposited near 115,000 BP, indicates a complex pattern of early Wisconsin crustal deformation associated with the growth of early Wisconsin ice.

Similarly, Sutherland (1981) accounted for the occurrence of high-level glaciomarine deposits in Scotland by suggesting that, during the period of rapid northern hemisphere ice accumulation at 75,000 BP, the Scottish ice sheet developed more rapidly than the large Laurentide and Eurasian ice sheets. Consequently, glacio-isostatic depression took place during a relatively early period of northern hemisphere glaciation and therefore enabled the deposition of marine sediments at relatively high levels prior to (ice-sheet-induced) global sea level lowering. Similar models could be applied to other Late Quaternary ice sheets (e.g. the Patagonian and New Zealand ice sheets) that were produced during the Early Wisconsin/Weichselian.

SEA LEVEL CHANGES DURING THE LAST 18,000 YEARS

Data from areas affected by glacio-isostatic uplift

Fennoscandia

The many detailed studies of raised shorelines in Fennoscandia have resulted in the production of shoreline diagrams in which the glacio-isostatic deformation of individual shorelines is depicted and where the chronological relationships between individual shorelines are demonstrated (Tanner 1930; Marthinussen 1960, 1962; Hafsten 1983; Anundsen 1985) (Figure 12.5) (see also p. 206).

The general shape of the uplift isobases for the emerged shorelines in Fenno-scandia shows that the greatest land uplift, and hence the greatest glacio-isostatic depression, took place over the northern Gulf of Bothnia (Figure 12.5). The elliptic nature of the shoreline isobases also indicates that the Fennoscandian dome of uplift functioned separately from the British Isles dome of uplift. Comparison of the uplift isobases with the reconstructed ice sheet profiles across Fennoscandia suggests that shorelines of similar ages have steeper regional gradients in western Norway where the former ice profile was steeper and more gently sloping gradients in the east where the ice profile had a more subdued slope (Anundsen 1985) (Figure 12.5).

Mörner (1971) attempted to calculate the pattern of eustatic sea level changes for Scandinavia using data on crustal uplift and relative sea level change. Calculation of crustal deformation was undertaken based on information on the regional gradients of raised shorelines of known ages. These enabled the amount of isostatic uplift to be calculated using shoreline time-gradient curves that, in turn, enabled a regional eustatic curve for the last 18,000 years to be calculated (Figure 12.6).

From these calculations, Mörner (1971) concluded that eustatic sea level in NW Europe during the last glacial maximum was between −85 and −90 m and that it subsequently rose to −60 m by 15,000 BP. Thereafter, eustatic sea level lowering took place until after 13,000 BP when there was a rapid rise (Mörner 1980a) until around 11,000 BP when sea level had reached −45 m. According to Mörner (1980a) sea level fluctuated slightly during the Younger Dryas between 11,000 and 10,000 BP. Thereafter it began to rise soon after 10,000 BP with the most rapid rise having occurred between 9,000 and 7,000 BP.

In southern Scandinavia, the complex interplay between glacio-isostatic uplift and eustatic sea level changes led to dramatic changes in geography. For example, following the drainage of the Baltic Ice Lake, a relative marine trans-gression resulted in the extension of marine waters eastwards into southern

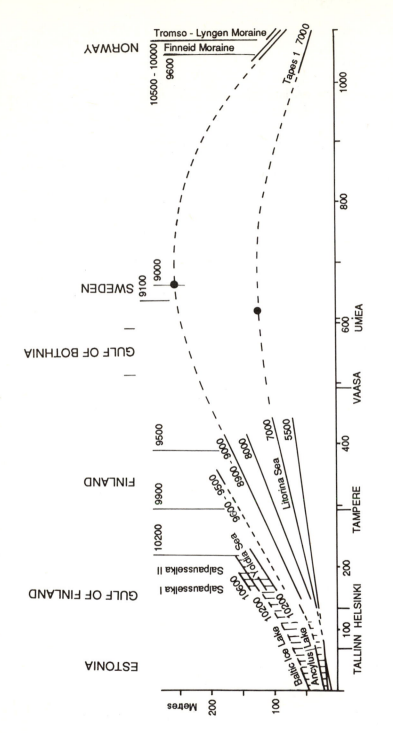

Figure 12.5 Shoreline diagram across Fennoscandia. The estimated altitudes (lines and dots) are connected with broken lines to show the assumed dome-like pattern of uplift. The positions of the water level of the Baltic Ice Lake and Ancylus Lake above sea level are shown with vertical lines. Some dated ice margin positions are also given (after Donner 1980).

Finland and the development of the Yoldia Sea (Figure 5.3B) (Eronen 1983). During the early Holocene, as glacio-isostatic uplift outpaced the eustatic rise in sea level, the Yoldia Sea became subject to relative marine regression and, by 9,500 BP, glacio-isostatic uplift over southern Sweden had resulted in the re-establishment of a freshwater lake in eastern Scandinavia although this time it was not ice dammed. By 9,000 BP this lake (the Ancylus Lake) had begun to decrease in size due to the creation of a new lake outlet in the Danish Sound (Eronen 1983). Thereafter, a new period of relative marine transgression over eastern and southern Scandinavia commenced at 8,500 BP as the continued rise of world sea level resulted in the development of the Litorina Sea in the southern Baltic. The culmination of this relative marine transgression took place between 7,500 and 7,000 BP. Thereafter, continued glacio-isostatic uplift over Scandi-navia resulted in the relative regression of the Litorina Sea.

The chronology of relative sea level changes in Scandinavia is additionally complicated since, due to differential glacio-isostatic uplift and sea level change, many of the relative sea level changes associated with specific transgressions or regressions took place at different times in different places (i.e. they are time-transgressive). For example, the beginning of the Litorina Sea in southern Finland

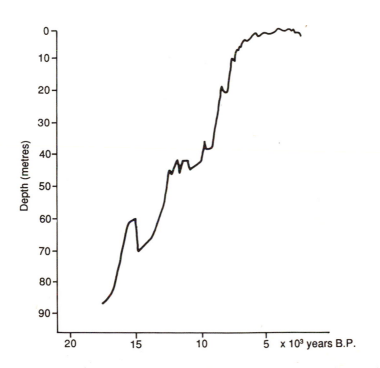

Figure 12.6 Eustatic changes in sea level across northwest Europe/northeast Atlantic region during the last 18,000 years as established from south Scandinavian data (after Mörner 1971).

is dated at 7,300–7,400 BP while in the northern Gulf of Bothnia it may have occurred nearer 7,000 BP (Hyvarinen 1973, Eronen 1974, 1983).

North America

Investigations of the relationships between former ice limits and relative shoreline displacements associated with the decay of the Laurentide ice sheet were mostly pioneered by Andrews (1970). Andrews' (1970) early studies of relative sea level change in Arctic Canada showed that the pattern of glacio-isostatic uplift did not correspond to that of a simple ellipsoid (Figure 12.7). Instead there appeared to

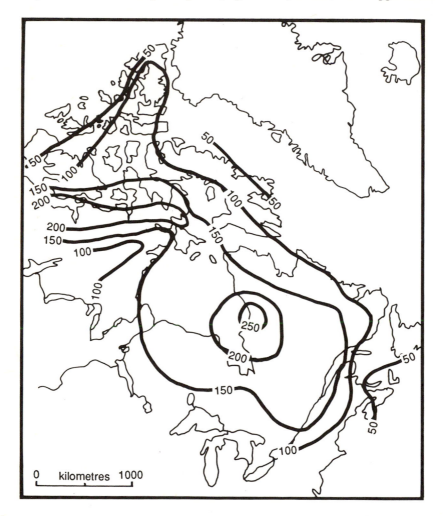

Figure 12.7 Distribution of the marine limit elevations (m) in Canada showing a complex pattern of glacio-isostatic uplift (after Andrews 1970).

be three distinct centres of glacio-isostatic uplift, each with a separate pattern of diminishing uplift away from the centres of former ice domes. The major centre of uplift was located over Richmond Gulf, SE Hudson Bay, with secondary uplift centres over the Keewatin District and Ellesmere Island.

In general, eustatic sea level around the margins of the Laurentide ice sheet appears to have risen from a minimum value of near −100 m (Hill *et al.* 1985) although the relative local values were much higher due to glacio-isostatic depression. Between 18,000 BP and 14,000 BP, the rapid rate of glacio-eustatic rise appears to have outpaced the rate of glacio-isostatic uplift of the land surface. Certainly, by 14,000 BP, sea level may have risen across the Scotia Shelf to −60 m (Piper *et al.* 1986), a value broadly similar to that estimated by Mörner (1971) for NW Europe and by Hill *et al.* (1985) for the Canadian Beaufort Shelf. Eustatic sea level fluctuations close to the Laurentide ice sheet between 14,000 and 10,000 BP are unclear although there is some support for the view that sea level had reached approximately −35 m by 10,000 BP over the Atlantic continental shelf. However, there were considerable regional variations in the position of relative sea level due to the effects of glacio-isostatic uplift. For example, relative sea level in the Ottawa valley had reached nearly +170 m by 11,000 BP (Dyke and Prest 1987). Similarly, by 9,000 BP, relative sea level along the northern margin of the Keewatin ice dome ranged between +240 m and +150 m.

One of the most widespread marine inundations took place shortly after 8,000 BP following the disintegration of ice in the Hudson Basin. At this time, the rate of glacio-eustatic sea level rise exceeded the rate of land uplift in the Hudson Basin and resulted in a relative marine transgression. The invading marine waters, known as the Tyrrell Sea, led to the widespread formation of shoreline features that have since been subject to considerable glacio-isostatic uplift. Thus despite 'eustatic' sea level at this time being between −20 m and −30 m, the highest raised shoreline of this age occurs at +315 m in the Richmond Gulf (Fairbridge 1983).

Scotland

A large amount of information on former sea level changes for this area has been derived through the detailed studies of Sissons and co-workers (e.g. Sissons 1967, 1974a, 1983; Sissons *et al.* 1966; Gray 1978; Dawson 1984). These, and other, studies have shown that, owing to the glacio-isostatic uplift caused by the relatively restricted size of the last ice sheet in this region, Lateglacial raised shorelines produced during regional deglaciation do not occur above +40 m OD (Ordnance Datum) (Plate 16). Similarly, the highest raised shoreline features produced during the Younger Dryas when regional eustatic sea level was at circa −45 m (Mörner 1971) are between +10 and +11 m (Sissons 1974b; Gray 1978). The presence of these shoreline features relatively close to the centre of glacio-isostatic uplift in the western Highlands therefore demonstrates that approximately 55 m of glacio-isostatic uplift have taken place in this region during the last 10,000 years. One important point to emerge from these studies has been provided by Gray (1974, 1978) who has shown that in parts of western Scotland, the Younger Dryas shoreline has been disrupted by faulting. This observation raises the possibility that emerged shorelines in other so-called 'stable' continental regions, where the rate of land uplift has been much faster, may have been similarly disturbed by tectonic activity.

Data from areas of relative tectonic stability

An extremely detailed curve of regional eustatic sea level change for the Caribbean and western North Atlantic has recently been provided by Fairbanks (1989) (Figure 12.8). Fairbanks (1989) used dated samples of the coral species *Acropora palmata* from Barbados which presently grows within 5 m of the sea surface. His sea level curve is for the last 17,000 years and is corrected for an assumed mean uplift of 0.34 m per thousand years. It indicates that sea level was at -121 ± 5 m during the last glacial maximum and that it subsequently rose at an accelerating rate until the beginning of the Younger Dryas (near 11,000 BP). Thereafter, there was a substantial decrease in meltwater discharge into the world's oceans until 10,000 BP. After this date, rapid sea level rise commenced again until it started to decelerate near 7,000 BP.

Fairbanks (1989) also attempted to relate the Barbados sea level curve to the deep sea oxygen isotope record. He showed that the principal ^{18}O minima coincided with time periods identified from his sea level curve when sea level was rising very rapidly. He concluded that the most rapid rates of sea level rise were centred on 12,000 BP and 9,500 BP when average meltwater discharge into the North Atlantic took place respectively at 14,000 km³ per year and 9,500 km³ per year (Figure 12.8B). During the former period, eustatic sea level appears to have risen by 24 m in less than 1,000 years. By contrast, sea level may not have risen by more than 20 m between 17,000 and 12,500 BP.

Fairbanks (1989) also observed that meltwater discharge during the Younger Dryas was remarkably low. For example, he concluded that, between 11,000 and 10,500 BP, the average meltwater discharge was only 2,700 km³ per year (the present discharge rates for the Mississippi and St Lawrence Rivers are 560 km³ per year and 330 km³ per year respectively) (Figure 12.8B). Accordingly, Fairbanks' sea level curve shows that sea level rose only slowly during the Younger Dryas, a view previously suggested by Mörner (1971) and probably due to the renewed growth of ice sheets.

SUMMARY

It has proved very difficult to measure and interpret patterns of sea level changes for the Late Quaternary. There are several reasons for this. First, in formerly glaciated environments, the observed patterns of sea level change may have been complicated due to the geoidal attraction of ocean water to ice sheets. Second, the pattern of global geoidal sea surface change is not known for the Late

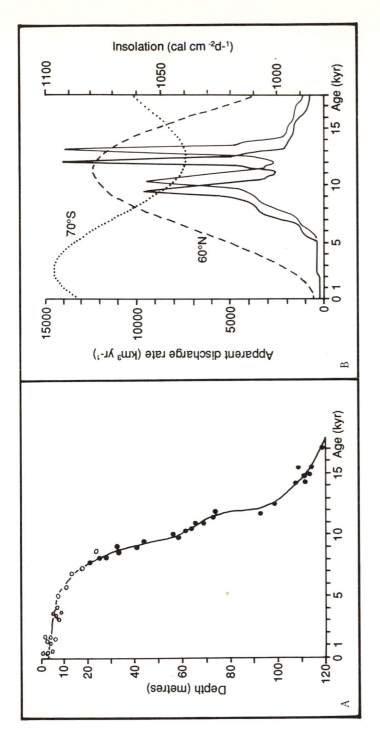

Figure 12.8 A. Barbados sea level curve for the last 18,000 years based on dating of submerged corals. B. The inferred rate of glacial meltwater discharge (km³ yr⁻¹) into the world's oceans calculated from the Barbados sea level data. Two different age estimations are given in ¹⁴C years (left) and calendar years (right). Milankovitch insolation curves for 60°N and 70°S are also shown (after Fairbanks 1989).

Quaternary. Third, it has proved very difficult to disentangle the effects of Late Quaternary hydro-isostatic and tectono-eustatic changes from those normally attributable to glacio-eustatic and glacio-isostatic effects. Fourth, it has been recognised that there is no such thing as a global eustatic sea level curve – such curves may have no more than regional significance. Finally, it has become increasingly clear that most coastlines (even along 'stable' trailing continental plate margins) have been subject to complex tectonic activity the details of which we are still unaware.

Despite these problems, Shackleton's (1987) curve of sea level change based on oxygen isotope data has great significance. Unlike other sea level curves, it has a measure of validity since it describes changes in the global volume of sea water contained within the world's oceans. By definition, it correlates closely with Late Quaternary global fluctuations in ice volume. The data also correlate well with the pattern of Late Quaternary sea level changes based on data from New Guinea (Chappell and Shackleton 1986). These curves show that global sea level may have fallen by as much as 50 m by isotope substage 5d and may have fallen as low as −75 m by the beginning of isotope stage 4 and to near −120 m during the last glacial maximum. Equally marked on the curves are the pronounced sea level rises that took place during the Late Quaternary. For example, global sea level rises in excess of +25 m took place during isotope substage 5c, 5a as well as on several occasions during stages 4 and 3 and presumably were related to the melting of large ice sheets.

The most detailed information for eustatic sea level changes for the last 18,000 years is provided by Mörner (1971) and Fairbanks (1989). Mörner's (1971) regional eustatic curve was calculated by removing the glacio-isostatic component from curves of relative sea level change. Fairbanks' (1989) detailed curve for the Caribbean is based on dated submerged coral and assumes an average rate of local tectonic uplift. Fairbanks has also used oxygen isotope data to calculate fluctuations in the rate of meltwater discharge from the northern hemisphere ice sheets into the North Atlantic. He has shown that ice sheet melting seems to have taken place in a stepwise manner with pulses of high discharge and rapid sea level rise (e.g. at 12,000 BP and 9,500 BP) having been separated by a period of low discharge and reduced rates of sea level rise (e.g. between 17,000 BP and 12,500 BP as well as during the Younger Dryas). The final deceleration in the rate of global sea level rise only appears to have commenced near 7,000 BP once most of the northern hemisphere ice sheets had finally melted.

NOTES

1   The relative sea level change at any location remains zero when the rate of absolute (eustatic) sea level change is exactly balanced by a corresponding rate of (isostatic) land uplift or subsidence. Any imbalance between these two rates of vertical movement will result in either a relative marine transgression or a relative marine regression.

13

MILANKOVITCH CYCLES AND LATE QUATERNARY CLIMATE CHANGE

MILANKOVITCH CYCLES

Introduction

In recent years, much attention has been paid to the view that the alternating cold and warm periods of the Quaternary were principally due to changes in the nature of the Earth's orbit around the Sun. This theory was first advocated by James Croll in the late nineteenth century (Croll 1867a,b, 1875) and later elaborated by Milutin Milankovitch (1941). In recent years, the mathematical basis of this astronomic theory of the ice ages has been substantially refined by Vernekar (1972) and by Berger (1978a,b, 1979). In particular, Berger (1978a,b, 1979) has provided highly detailed information on past insolation variations.

One important assumption of the Milankovitch theory of orbital changes is that there has been no absolute annual change in the amount of incoming solar radiation. Long-term changes in the Earth's orbit are believed to cause a redistribution of insolation across both hemispheres. It is very likely that the occurrence of flares on the Sun's surface has altered the absolute amount of solar radiation received at the outer atmosphere throughout the Late Quaternary. Needless to say, however, it is impossible to estimate the magnitude of such long-term changes!

In the following pages, an account is given of the timing and duration of Late Quaternary global environmental changes and possible relationships with Milankovitch orbital processes. This is a controversial topic because the identification of particular warming and cooling episodes is largely dependent on accurate age determinations of geological events. However, it is well known that age estimates of particular events are often of uncertain reliability. Accordingly, in this discussion of Late Quaternary climate changes, use is made of published ages that are considered most representative of particular environmental changes. It is also assumed that the timescales for the Late Quaternary based on Milankovitch astronomical data, oxygen isotope stratigraphy and radiometric age determinations are broadly comparable (Martinson *et al.* 1987).

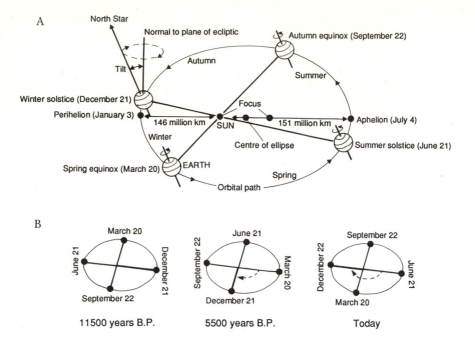

Figure 13.1 A. Geometry of the Sun–Earth system. The Earth's orbit, an ellipse with the Sun at one focus, defines the plane of the ecliptic. The Earth moves around its orbit in the direction of the arrows, while spinning about its own axis, which is tilted to the plane of the ecliptic at circa 23.5° and points toward the North Star. B. Precession of the equinoxes causes the position of the equinoxes and the solstices to shift slowly around the Earth's elliptical orbit. During the Lateglacial, the winter solstice occurred near aphelion whereas today it occurs at the opposite end of the orbit near perihelion (after Skinner and Porter 1987).

Milankovitch orbital processes

At present, the Earth has an elliptical orbit around the Sun. The Sun is not located at the centre of the ellipse but instead occurs at one of the foci (Figure 13.1). During the winter solstice in the northern hemisphere, the Earth is at one end of its elliptical orbit when it is nearer the Sun (in perihelion) and thus receives greater heat. In contrast, the northern hemisphere summer solstice is character-ised by an elliptical orbit position more distant from the Sun (in aphelion). The principal cause of the seasons is the tilting of the Earth's axis at an angle that is not perpendicular to the plane of its orbit. During the course of an annual Earth orbit, the tilting of the Earth results in the seasonal heating of each hemisphere. The timing of perihelion presently coincides with the period when the southern hemisphere is tilted towards the Sun. Conversely, aphelion coincides with the northern hemisphere summer when the Earth receives approximately 3.5 per cent less solar radiation than the annual mean (as measured at the outer edge of the Earth's atmosphere).

Throughout time, there has been a continual change in the eccentricity of the elliptical orbit. Marked changes in the axial tilt (obliquity) of the Earth have also taken place. In addition, there also have been changes in the timing of perihelion and aphelion with respect to seasonal changes on Earth (the precession of the equinoxes).

Changes in orbital eccentricity

Changes in orbital eccentricity have varied with time from a circular orbit (when perihelion and aphelion are identical) to maximum eccentricity when the values of incoming solar radiation may have varied by as much as 30 per cent between perihelion and aphelion. The periodicity of this cycle is 95,800 years during which time the Earth alternates from a circular orbit to a highly eccentric orbit and back again to a circular orbit. It should be stressed that changes in orbital eccentricity do not cause any change in the amount of solar radiation reaching the Earth during summer or winter nor any change in the total annual heat received by either hemisphere. Instead, the effect is to increase the contrast in seasonality in one hemisphere and reduce it in the other. Croll believed that when this contrast is at its maximum, it would cause increased snowfall in the northern hemisphere during winter. The increased global albedo due to a widespread snow cover might modify the climate of the succeeding seasons and, in this way, initiate widespread glaciation.

Changes in inclination (obliquity)

Changes in the inclination of the Earth's axis have varied between extreme values of 21.39° and 24.36° (the present value is 23.44°) with a periodicity of 41,000 years. Increases in axial tilt result in lengthening of the period of winter darkness in polar regions. They also result in changes in the seasonal latitudinal range in which the Sun occurs overhead. Changes in obliquity therefore cause significant changes in the receipt of solar radiation at high latitudes but do not greatly affect the amount of incoming solar radiation at low latitudes. Owing to equal changes in axial tilt in both hemispheres, changes in incoming solar radiation are the same in both hemispheres.

Precession of the equinoxes

Variations in the timing of perihelion and aphelion are caused by a 'wobbling' in the Earth's axis of rotation as it rotates around the Sun. Over a period of 21,700 years, the axis slowly swings in a conical manner around a line perpendicular to the orbital plane (Figure 13.1). Over this time period, the northern hemisphere is tilted towards the Sun at successively different points on the Earth's orbit. At

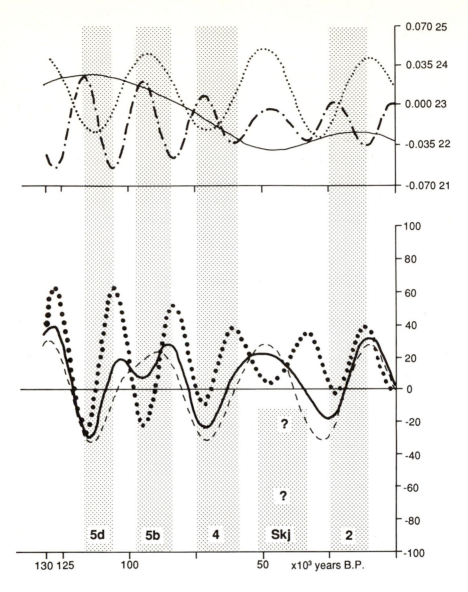

Figure 13.2 Variations in eccentricity (dashed line), precession (stippled line) and obliquity (continuous line) for the last 130,000 years (top diagram) and northern hemisphere summer solar radiation values for 80°N (continuous line), 65°N (dotted line) and 10°N (stippled line) expressed as departures from 1950 AD values (lower diagram). For upper diagram scale bar information, see Berger (1978a, b). Also shown are the principal periods of Late Quaternary glaciation as defined from oxygen isotope stratigraphy and geological data (various sources). An additional period of Scandinavian glaciation, the Skjonghelleren Stadial (Skj), is also shown for part of oxygen isotope stage 3 (after Berger 1978a, b, 1979; Bradley 1985; Larsen and Sejrup 1990).

present, this takes place during summer when the Earth is in aphelion. However, approximately 11,000 years ago during the Younger Dryas, the Earth was tilted towards the Sun in mid-summer during perihelion. In theory, therefore, northern hemisphere winters during the Younger Dryas were colder and longer than present while summers were shorter and warmer.

Combined effects

The solar radiation received in low latitude areas is principally affected by variations in eccentricity and precession of the equinoxes. By contrast, higher latitudes are mainly affected by changes in axial tilt (obliquity). The combined influence of changes in eccentricity, obliquity and precession of the equinoxes produces a complex pattern of insolation variations (Figure 13.2). These are indicative of increased seasonal contrasts in one hemisphere and diminished contrasts in the other. However, it is not known if, during certain time periods in the Late Quaternary, insolation variations in the high latitudes of the northern hemisphere may have induced similar environmental changes in the southern hemisphere. Similarly, it is not known if, on certain occasions, insolation changes in the high latitudes of the southern hemisphere (particularly in Antarctica) may have led to a climatic response in the northern hemisphere. Indeed some have argued that climate changes in both hemispheres have taken place in an approximately synchronous manner (Broecker and Denton 1990a,b).

A critical factor that affected rates of ice sheet build-up and decay in the middle latitudes during the Late Quaternary was the seasonal temperature gradient between low and high latitudes – known as the insolation gradient (Young and Bradley 1984). During periods when the insolation gradient from the equator to the pole was high, meridional (north–south) circulation would have been increased, thus providing a greater rate of delivery of snow-bearing precipitation to high latitudes. By contrast, low meridional temperature gradients would have been associated with oscillatory jet stream flow, the more frequent development of blocking high pressure systems in the middle latitudes and a decrease in the rate of supply of moisture to high latitudes.

Milankovitch cycles and climate change

It has often been argued that the observed patterns of Late Quaternary environmental changes reflect well-defined cycles of Milankovitch insolation. For example, it is generally agreed that the timing of the principal glacial–interglacial cycles during the Quaternary reflects the influence of the 96,000 year eccentricity cycle (Imbrie and Imbrie 1979) and that the last glacial–interglacial cycle, approximately corresponding to the Late Quaternary, was principally due to this process. For a shorter timescale, Mason (1976) proposed that Late Quaternary environmental changes were indicative of the occurrence of the 41,000 year Milankovitch obliquity cycle. Similarly, attention has been drawn to the likelihood that the principal Late Quaternary environmental changes that took place may have been due to a 21,000 year precessional cycle (e.g. Chappell 1974c, Larsen and Sejrup 1990) (Figures 13.2 and 13.6). Although the majority of Late Quaternary time lies within a single Milankovitch eccentricity cycle of circa 100,000 years, a series of precessional and obliquity cycles took place during this time period. Consideration is given here to possible relationships between the timing of these cycles and other parameters of Milankovitch insolation with the most important Late Quaternary climate changes.

Insolation variations and global climate change

The early calculations of Milankovitch provided information on the variations in incident solar radiation, as a function of season and latitude, for the last million years. Particular attention was given to the reconstruction of insolation for the high latitude regions of the northern hemisphere since these areas are bordered by numerous mountain ranges with high snowfall and are thus likely to have been sensitive to the former growth and decay of glaciers and ice sheets. More recently, Berger (1978b, 1979) has calculated past variations in midmonth daily insolation for each 10° latitude at intervals of 1,000 years (Figure 13.3). He also computed long-term differences (anomalies) in insolation from the mean state of midmonth daily insolation for the critical latitude 60°N (Figure 13.4). In this way, he identified, for intervals of 1,000 years, particular time periods throughout the year when the reconstructed insolation was higher than average (insolation maxima) and lower than average (insolation minima).

The numerical analyses of Berger (1978a,b, 1979) show that a characteristic feature of Late Quaternary insolation variations has been a forward shift of an insolation maximum through the summer months from May towards August. At the same time, an insolation minimum begins to develop in February and thereafter drifts towards early summer. This feature, known as the insolation signature, is considered by Berger (1979) to be related to changes from global warmth to the onset of colder conditions. At certain times, a spring insolation minimum is rapidly replaced by an insolation maximum that shifts towards the summer months. Berger argued that during such periods, global climate may change from cold to warm.

However, there are other methods by which seasonal changes in Milankovitch insolation may cause widespread climatic change. For example, Ruddiman and McIntyre (1981c) maintained that global ice growth is favoured during periods when summer insolation is low (leading to decreased ice ablation) and winter insolation is high (leading to increased moisture evaporation from oceans and increased snow supply to adjacent continents). Conversely ice decay may be favoured by high summer insolation (promoting increased ice ablation) and low winter insolation (decreased moisture evaporation from oceans).

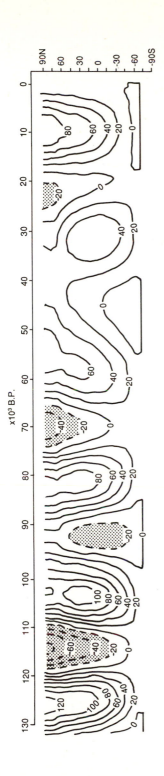

Figure 13.3 Midmonth values of daily insolation for July in thousand year intervals before 1950 AD. The shaded areas indicate areas where the values are negative (after Berger 1978a, b, 1979). Note the relatively rapid changes in midmonth insolation between 130,000 and circa 65,000 BP and the relatively small changes between circa 65,000 BP and the present (see also Figure 13.6).

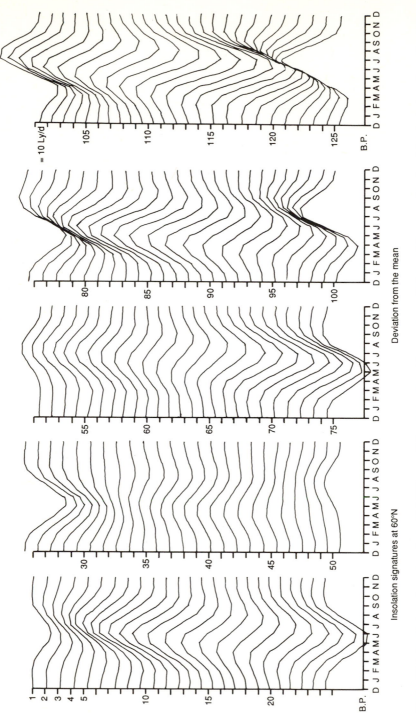

Figure 13.4 Insolation signatures from 125,000 years BP to present and long-term variations of the annual cycle of the deviations from the mean state of the midmonth daily insolation for 60°N. The vertical scale is indicated (Langleys per day). Note the way in which there are long-term seasonal drifts of individual insolation minima and maxima (after Berger 1979).

PATTERNS OF LATE QUATERNARY CLIMATE CHANGE BASED ON GEOLOGICAL AND MILANKOVITCH CRITERIA

Introduction

In the following section, an attempt is made to identify the principal periods of relative global warmth as indicated by geological evidence and to discover if there is any correspondence between the timing of these events and periods of Milankovitch warming. This is followed by a summary of the field evidence used to indicate the principal periods of global cooling that have taken place throughout the Late Quaternary.

Periods of relative warmth

Warming during substage 5e

The beginning of the Late Quaternary, corresponding to oxygen isotope substage 5e, was characterised by exceptional global warmth. This time period, considered to have lasted between approximately 130,000 and 122,000 BP, may have been associated with sea levels up to 6 m higher than present. This view is consistent with the oxygen isotope record which shows that substage 5e corresponds with a clearly defined peak in most ocean cores that indicates a global ice volume less than at the present. One explanation for the reduced ice volume is that the West Antarctic ice sheet may have disintegrated during this time period. Other views include widespread thinning of parts of the East Antarctic ice sheet at this time as well as the possible disappearance of the southern dome of the Greenland ice sheet (Funder 1989). This non-glacial interval is represented in many continental areas by well-defined palaeosols, for example in the loess areas of China and the USSR.

This period of pronounced warmth appears to coincide with a period of strong Milankovitch global warming centred on 125,000 BP and lasting between approximately 134,000 BP and 120,000 BP when solar insolation values were significantly higher than present. There is a clear correspondence between the inferred peak of last interglacial warmth at 127,000 BP and the strongest positive radiation anomaly of the Late Quaternary. This period of interglacial warmth may also have been characterised by weak meridional circulation (Young and Bradley 1984).

Warming during substages 5c and 5a

The oxygen isotope record shows that during stage 5 two periods of interstadial warmth occurred after the exceptional interglacial warmth of substage 5e. The older of these corresponds with substage 5c with the younger occurring during substage 5a. According to the oxygen isotope curve of Martinson *et al.* (1987)

(see Chapter 1) substage 5c occurred between 105,000 BP and 93,000 BP while substage 5a represents the time interval between 85,000 BP and 74,000 BP. In Europe, these interstadials may correspond to the Brørup and Odderade events respectively. Both interstadials are well defined on the oxygen isotope record and, according to Shackleton's (1987) corrected curve, both time intervals coincide with significant rises in global sea level – perhaps of the order of 20–30 m (Figure 12.4).

Several authors have suggested that both warming episodes are represented by emerged coral reef terraces indicative of relative marine transgressions. For example, it has been argued that the Barbados II high sea level interval may have occurred near 105,000 BP and possibly represents substage 5c (Mesolella *et al.* 1969). Similar conclusions have been inferred from the sea level data from Papua and New Guinea (Chappell 1974a; Chappell 1987). In Arctic Canada, Andrews *et al.* (1986) have suggested that a pronounced period of ice recession took place during substage 5c. The Barbados sea level data indicate the occurrence of a second period of relatively high sea level (Barbados I) during isotope substage 5a and tentatively dated at 82,000 BP.

There is a surprisingly good correspondence between these inferred periods of interstadial warmth and the periods of warmth suggested by the Milankovitch data. For example, a strong positive insolation departure during substage 5c was centred on 105,000 BP and extended between 108,000 BP and 97,000 BP. Similarly, a prolonged period of increased insolation (above 1950 AD values) occurred between 90,000 BP and 77,000 BP and was centred on 83,000 BP. Both of these isotope substage time intervals also coincide with periods of increased orbital precession (Figure 13.2).

Interstadial warmth during isotope stages 4 and 3

Isotope stages 4 and 3 are generally considered to correspond respectively with the time intervals between 74,000 and 59,000 BP (stage 4) and between 59,000 and 24,000 BP (stage 3) (Martinson *et al.* 1987). According to the chronology of oxygen isotope stages, stage 4 corresponds to a period of global ice accumulation while stage 3 coincides with a period of global warming. However, the pattern of climate changes during stages 4 and 3 appears to have been very complex with at least three periods of interstadial warmth having taken place.

(1) The oldest period of interstadial warming appears to have occurred near the end of stage 4 at about 60,000 BP. For example, evidence from palaeosols in eastern Europe and western USSR implies the occurrence of a period of temperate warmth at this time. Supportive evidence for a period of global warmth between approximately 60,000 and 50,000 BP is provided from several interstadial sites in NW Europe and Scandinavia. For example, in western Norway, the Bø interstadial is thought to have taken place approximately 60,000 years ago (Larsen and Sejrup 1990). Biostratigraphical evidence from Grand Pile is consistent with the view that a well-defined period of interstadial warmth (St

Germain II) occurred at this time (Mook and Woillard 1982). Similarly, in North America climatic amelioration is also believed to have taken place shortly after 65,000 BP (Port Talbot interstadial I).

During this time interval, widespread retreat and decay of Laurentide and Cordilleran ice is thought to have occurred. Clague (1981) and Fulton *et al.* (1986a) have argued that the climatic warming may also have caused the virtual disappearance of the Cordilleran ice during this period. Similarly, Lundqvist (1986b) has drawn attention to the widespread climatic amelioration that took place in Scandinavia during this period and has argued that the virtual complete deglaciation of Early Weichselian ice in Scandinavia may have taken place at this time.

Although it is possible that widespread ice retreat and decay may have occurred throughout the northern hemisphere between 60,000 and 50,000 BP, the absence of numerous interstadial sites of this age in North America suggests that the evidence for widespread deglaciation at this time is not entirely unequivocal. For example, the periglacial evidence from European USSR does not indicate global warming for this period. Thus, Velichko and Nechayev (1984) consider that a major period of permafrost development took place between 50,000 and 60,000 years ago. By contrast, Baulin and Danilova (1984) consider that a marked period of climatic warming and soil development on the Russian Plain took place slightly earlier between 65,000 and 70,000 BP. A compelling argument, however, for significant global warming following 60,000 BP is provided by Shackleton (1987). His curve of Late Quaternary sea level change based on oxygen isotope stratigraphy indicates that there may have been a substantial eustatic sea level rise (between 30 and 40 m) between circa 60,000 and 55,000 BP.

(2) There is equivocal evidence for a period of climatic warming during isotope stage 3 between 50,000 and 35,000 BP. For example, in Denmark, organic sediments dated to approximately 37,000 BP (the Hengelo interstadial) may indicate a period of climatic amelioration (Houmark-Nielsen and Kolstrup 1981). Similarly, in Britain interstadial conditions may have existed between approximately 43,000 and 40,000 BP (the Upton Warren interstadial) (Mitchell *et al.* 1973; Shotton 1986) while interstadial conditions may also have existed in part of North America between 40,000 and 47,000 BP (Port Talbot interstadial II) (Dredge and Thorleifson 1987).

There is contradictory evidence, however. For example, a major ice sheet expansion (the Skjonghelleren Stadial) took place in Scandinavia between approximately 55,000 BP and 33,000 BP with a particular severe episode of glaciation (the Jaeren Stadial) having occurred between approximately 39,000 and 41,000 BP (Andersen *et al.* 1981, 1987; Larsen and Sejrup 1990). Similarly, Alam and Piper (1977) have argued that a drastic cooling took place in Atlantic Canada between approximately 40,000 and 30,000 BP. However, Shackleton's (1987) oxygen isotope data imply that global ice volumes were shrinking at this time since a prominent interval of global sea level rise is centred on 40,000 BP.

(3) A marked warming interval appears to have taken place near the end of isotope stage 3 sometime between 33,000 and 25,000 BP. In European USSR, Velichko (1984) has shown that this period of warmth was associated with widespread soil development. Similarly, Sakaguchi (1978) has observed evidence from Japan for climatic warming for this time interval while in eastern Canada, the Plum Point interstadial is dated between 35,000 and 23,000 BP (Fulton et al. 1986b). In central Alaska, a relatively mild period of moist conditions associated with widespread thawing of ground ice took place between 27,000 and 25,000 BP (Mathews et al. 1989). Similarly, the levels of numerous African lakes appear to have been relatively high before falling during isotope stage 2 (Hamilton 1982) (see Chapter 7).

Interstadial sites of an approximately similar age occur in Europe. For example, the Ålesund interstadial of western Norway is dated between 33,000 and 28,000 BP (Larsen et al. 1987). In Denmark, a period of interstadial warmth (the Denekamp interstadial) is indicated for the time period between 32,000 and 29,000 BP (Kolstrup 1980) while in western Scotland, interstadial deposits dated between 30,000 and 29,000 BP may indicate a similar period of relative warmth (Jardine et al. 1988). Velichko and Nechayev (1984) have described interstadial conditions and related soil development in European Russia that took place between 29,000 and 24,000 BP. Similarly, Baulin and Danilova (1984) have argued that soil development was widespread throughout the USSR during this period and was associated with the widespread degradation of permafrost. In general, it appears that interstadial conditions may have lasted till 29,000 BP and possibly in some areas till as late as 25,000 BP. This is quite remarkable since it appears to indicate that the later growth of the last major ice sheets in North America and Eurasia may have taken place during a comparatively short time interval between 25,000 and 18,000 BP.

The reconstructions of Milankovitch insolation for these time periods show a remarkable uniformity in value between 65,000 and 30,000 BP (Figure 13.3). During this time period there were relatively few pronounced seasonal variations in midmonth insolation. For this reason Berger (1979) concluded that global climate remained cold throughout this interval with interstadials superimposed on this overall trend. Inspection of the reconstructed insolation signatures for this time period show that a weak May maximum at 62,000 BP shifts to July by 58,000 BP (Figure 13.4). Berger (1979) has suggested that this shift defines an interstadial centred on 59,000 BP. The insolation values exhibit a slight yet progressive decline between 55,000 BP and 30,000 BP. During this time interval, the insolation signatures remain relatively uniform with no marked seasonal departures. Yet during this period there were a series of marked climatic fluctuations superimposed upon an overall cooling trend. This is well illustrated from the Grand Pile pollen record where the time interval separating the St Germain II interstadial event and the last glacial maximum at 18,000 BP is characterised by a series of pronounced environmental changes (Figure 7.1). It is also indicated by the Dansgaard–Oeschger oscillations in the Greenland ice cores (see Chapter 2).

246

There is no clear correspondence between the interval of warming between approximately 33,000 and 28,000 BP and the patterns of Milankovitch insolation (Figures 13.2–13.4). For example, the pattern of insolation at 33,000 BP has a weak summer insolation maximum but also a weak spring insolation minimum. Thereafter there is a slight decline in spring and summer insolation that commences near 31,000 BP and becomes very pronounced by 25,000 BP (Figure 13.4).

Warming during the isotope stage 2/1 transition

Widespread global warming led to the melting of the last great ice sheets. The principal period of warming began near 15,000 BP and, with the exception of the Younger Dryas, continued uninterrupted into the Holocene interglacial. This is well illustrated by the glacial meltwater discharge rates calculated by Fairbanks (1989) (Figure 8.7B). These show that the period of maximum meltwater discharge took place between approximately 15,000 and 12,000 BP with a secondary peak between 10,000 and 9,000 BP. There is good agreement between this period of climatic warming and calculated values of Milankovitch insolation. The midmonth summer insolation anomalies became positive (insolation higher than present) soon after 19,000 BP and reached maximum values at 10,000 BP when they were as high as $+100$ cals/cm^2/day (Figure 13.3). During the early part of the Holocene interglacial, the midmonth July anomalies exhibit a progressive decline although they still remain positive.

Summary

In general there is a good agreement between the principal periods of global warming indicated by geological evidence and the inferred periods of climatic warming based on Milankovitch insolation variations. According to Ruddiman and McIntyre (1981c) high summer insolation and decreased winter insolation are likely to have provided conditions most suitable for climatic warming and ice decay. Certainly there is a close correspondence between periods of high summer insolation and reduced ice cover. Thus, apart from exceptional warmth during the last and present interglacials, the warmest interstadial conditions were centred on 105,000 BP and 82,000 BP with less pronounced interstadial conditions having occurred near 60,000 BP, 40,000 BP and 30,000 BP. With the exception of the interstadial event near 30,000 BP, all of the Late Quaternary interstadials correspond with time intervals characterised by strong positive summer insolation signatures. Possible correlations should be tempered with caution, however, since the calibration between Milankovitch astronomical chronology and radiometric ages is not known with certainty. There is additional reason for caution since it is not at all clear how Milankovitch effects are translated into changes in global climate.

Periods of global cooling

Isotope substages 5d and 5b

According to the oxygen isotope stratigraphy of Martinson *et al.* (1987), isotope substages 5d and 5b occurred respectively between 117,000 and 105,000 BP and between 93,000 and 85,000 BP. Richmond (1986a,b) has argued on several grounds that a period of glaciation took place in Wyoming during isotope substage 5d. Widespread ice advance is also thought to have taken place in Arctic Canada during this time period – the Ayr Lake stadial (Andrews *et al.* 1983, 1985, 1986). However, there is reason to believe that this stadial may incorporate a later period of ice advance that took place during substage 5b (Andrews and Miller 1984). In Scandinavia, two separate periods of ice advance (during the Gulstein and Eikelund stadials) may also have occurred at this time (Larsen and Sejrup 1990). Although there is at present only limited evidence from onshore data to suggest periods of ice expansion during substages 5d and 5b, the offshore data are less equivocal. For example, Shackleton (1987) has shown quite clearly that a major period of global ice build-up appears to have occurred during substage 5d while a less pronounced episode of ice expansion took place during substage 5b. The ice accumulation during substage 5d appears to have culminated at approximately 115,000 BP and may have been associated with a sea level lowering of over 40 m. The marked fall in sea level at this time is also recorded from the emerged coral reef sequences in New Guinea and Barbados (see Chapter 11). The occurrence of a marked sea level fall during substage 5b is less clear although a fall of nearly 10 m may have taken place (Shackleton 1987).

There is good correspondence between the pattern of Milankovitch insolation signatures and the episode of global cooling centred on 115,000 BP (Figures 13.3 and 13.4). This time period was characterised by extremely low midmonth insolation anomalies (as low as nearly -90 cals/cm^2/day), lower than at any other time during the Late Quaternary. It was also a time period characterised by substantially reduced summer insolation (Figure 13.4) and strong meridional (north-south) insolation gradients that may have persisted until near 110,000 BP (Young and Bradley 1984). A later and relatively weak summer insolation minimum centred on 93,000 BP may account for the pattern of global cooling during isotope substage 5b. However, there are other explanations.

The occurrence of widespread cooling during substage 5b coincides with a 'spike' in the oxygen isotope stratigraphy from Camp Century identified by Johnsen *et al.* (1972) (Figure 10.2). The 'spike', indicative of a rapid ice accumulation, appears to have occurred soon after 90,000 BP (but possibly as late as 84,000 BP (Drexler *et al.* 1980)) and has been proposed by Kennett and Huddlestun (1972) as due to climate cooling as a result of volcanism (see Chapter 10).

Global cooling during the stage 5/4 transition

Ruddiman and McIntyre (1981c) argued that low summer insolation and high winter insolation favoured rapid ice sheet growth in the northern hemisphere during the isotope stage 5/4 transition. They considered that, during this period, moisture evaporation during winters from a 'warm' ocean resulted in rapid ice accumulation on adjacent 'cool' continents (see Chapter 3). The marked thermal contrasts along the margins of continents would have resulted in the movement of mid latitude cyclones towards the developing ice sheets, thus favouring rapid ice accumulation. They proposed a model of ice sheet growth for this time period in which the increases in ice volume lagged behind the Milankovitch summer insolation minimum and that these changes were followed by a gradual decrease in ocean temperatures (Figure 13.5).

It is generally agreed that one of the most well-defined periods of global ice accumulation on the oxygen isotope stratigraphy took place between the close of isotope substage 5a and the beginning of stage 4 near 75,000 BP (see Chapter 3). Between 80,000 BP and 70,000 BP there may have been a glacio-eustatic sea level lowering of nearly 50m (Shackleton 1987). Ruddiman *et al.* (1980) have shown using ocean sediment evidence that rapid northern hemisphere glaciation may have taken place during this period. Similarly, Larsen *et al.* (1987) and Larsen and Sejrup (1990) have shown that extensive glaciation of Scandinavia took place at this time (the Karmøy Stadial). A similar period of very extensive early Wisconsin glaciation is considered to have taken place in North America (Vincent and Prest 1987) that in some areas may even have been more extensive than during the Late Wisconsin. For example, Richmond and Fullerton (1986b: 191) have

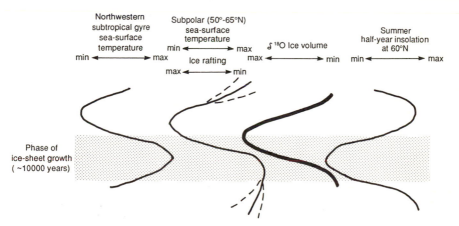

ICE-GROWTH SEQUENCE

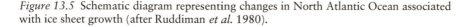

Figure 13.5 Schematic diagram representing changes in North Atlantic Ocean associated with ice sheet growth (after Ruddiman *et al.* 1980).

suggested that glaciation at this time was very extensive throughout east central and eastern North America while there may also have been widespread glacier build-up in the Coast and Cascade Ranges of NW USA and western Canada (Easterbrook 1976; Armstrong and Clague 1977). A different view has been adopted by Dredge and Thorleifson (1987) who maintain that the large areas of ice that existed in North America during isotope stage 4 may simply represent the expansion of ice masses that initially developed during substages 5b and 5d.

One of the most striking aspects of the oxygen isotope curves is the way in which the Earth's climate appears to have changed from dominantly interglacial to glacial conditions at approximately 75,000 BP. The computed patterns of insolation departures for this time period for 60°N show that incoming radiation was approximately similar to present at 75,000 BP but thereafter was low during the succeeding 10,000 years. However, by 75,000 BP, a well-defined summer insolation minimum was matched by a winter insolation maximum. This pattern remained essentially unaltered until 67,000 BP when the summer insolation minimum occurred in conjunction with a spring insolation maximum (Figures 13.3 and 13.4).

A factor that may be relevant to the rapid northern hemisphere ice sheet build-up at the stage 5/4 transition is the volcanic eruption at Toba in northern Sumatra (Ninkovich *et al.* 1978). The eruption is estimated by oxygen isotope stratigraphy and K–Ar dating to have occurred near 75,000 BP and is considered to have been the largest magnitude explosive eruption in the entire Quaternary. The timing and magnitude of the eruption during a marked summer insolation minimum raises the possibility that the eruption may have triggered and/or accelerated global cooling and the growth of ice sheets in the northern hemisphere. However, much more information on the precise age of the eruption and the timing of worldwide glaciation inception is needed before an answer can be provided to this question. The widespread glaciation that took place during the beginning of isotope stage 4 (irrespective of its cause) is also remarkable for its brevity since most of the ice that accumulated appears to have disappeared by 60,000 BP (see above).

Stage 3 cooling events

The extent of global glaciation during isotope stage 3 (between 59,000 and 24,000 BP according to Martinson *et al.* 1987) is not known with any certainty. Larsen and Sejrup (1990) have argued that the outer coast of western Norway (and hence most of Scandinavia) was covered by ice between 60,000 BP and 30,000 BP during the Skjonghelleren Stadial. According to Shackleton (1987), global sea level remained low (mostly between −50 m and −80 m) throughout most of this period although significant fluctuations of the order of ±20 m appear to have taken place. The evidence from North America is more equivocal. For example, Gascoyne *et al.* (1980) have argued that large areas of western Canada were not glaciated between 67,000 BP and 28,000 BP. The prevailing view, however, is that glaciation was widespread in North America during this period (Fulton *et al.* 1986a).

Recognition of the principal phases of ice sheet growth during stage 3 is additionally complicated by several suggestions that major ice expansion took place between 40,000 and 30,000 BP. For example, Kind (1975) considered that in Siberia a drastic climatic deterioration associated with widespread glaciation (the Zigansk stage) took place between 33,000 and 30,000 years ago and was followed by widespread permafrost degradation (Baulin and Danilova 1984). Similarly, Sakaguchi (1978) proposed that a period of very cold climate occurred in Japan at 31,000 BP while Alam and Piper (1977) have proposed that a dramatic cooling may have taken place in Atlantic Canada between approximately 40,000 and 30,000 BP. In western Norway, there seems to have been a very extensive glaciation centred on 40,000 BP (Andersen *et al.* 1987; Larsen and Sejrup 1990). Some of the climate cooling events described above may also be indicated by the Dansgaard–Oeschger oscillations evident in the Greenland ice core record. Although the climatic significance of these cooling 'spikes' is not yet understood, the possibility exists that numerous very severe episodes of climate cooling took place in the northern hemisphere during isotope stage 3.

Inspection of the seasonal patterns of Milankovitch insolation for stage 3 does not provide many clues about the patterns of climatic change that may have taken place during this time period. The curves show that this interval was characterised by relatively minor variations in the amount of summer insolation and that the overall pattern of insolation was similar to present (summer insolation slightly higher and winter values slightly lower than present). Thus, it does not seem likely that the growth of ice sheets during isotope stage 3 was a simple result of Milankovitch insolation variations.

Stage 2 Cooling

The global cooling that led to the widespread growth of northern hemisphere ice sheets during the last glaciation began sometime between 28,000 BP and 25,000 BP and lasted until approximately 15,000 BP. In the USSR, climate cooling appears to have commenced at approximately 25,000 BP (Velichko 1984) while in northern Europe, the build-up of ice appears to have followed an interstadial that ended near 25,000 BP (Larsen and Sejrup 1990). In North America, a major advance of the Laurentide ice sheet appears to have taken place soon after 25,000 BP after the Plum Point interstadial, while the maximum ice advance in many areas may have occurred as early as 21,000–20,000 BP (Dreimanis *et al.* 1966; Dredge and Thorleifson 1987). Similarly, the inferred water levels of Lake Bonneville are considered to have begun to rise near 25,000 BP and to have continued to do so until near 17,000 BP (Smith and Street-Perrott 1983).

It is not clear if the northern hemisphere ice sheets built up at this time on unglaciated continents. The balance of the evidence suggests that they did not and it is likely that the growth of the Late Wisconsin Laurentide ice sheet was associated with the growth and expansion of pre-existing ice masses. Similar conditions may have prevailed in Fennoscandia although western Canada may have been an exception since here, Late Wisconsin Cordilleran glaciation may have commenced upon an ice-free land surface (Chapter 4). Considered together, the field evidence indicates that the build-up of the last northern hemisphere ice sheets took place relatively rapidly with most ice accumulation having occurred between approximately 25,000 and 20,000 BP. The Late Quaternary sea level curve based on Shackleton's (1987) data shows these changes quite clearly (Figure 12.4). Between 25,000 and 20,000 BP global sea level appears to have progressively fallen from near $-75\,$m to $-120\,$m, the latter value being in good agreement with that calculated by Fairbanks (1989) (see Chapter 12).

According to Berger (1979), a Milankovitch spring insolation minimum began to deepen rapidly at 27,000 BP and had moved to May–June by 22,000 BP (Figure 13.4). However, by 19,000 BP the summer minimum had significantly weakened and had been replaced by a spring insolation maximum. Throughout the time interval between 27,000 BP and 18,000 BP the negative July insolation anomalies were considerably weaker than they were during the stage 5/4 transition or during isotope substages 5b or 5d. Thus the growth of the northern hemisphere ice sheets during the last glaciation was associated with solar insolation variations that were not as extreme as during earlier time intervals during the Late Quaternary (Figure 13.2). However, the meridional (north–south) insolation gradients at approximately 23,000 BP were relatively high and may have been an important factor in causing widespread glaciation (Young and Bradley 1984).

Younger Dryas cooling

It is perhaps appropriate here to mention the Younger Dryas since the occurrence of this period of cold climate shows that widespread cooling (possibly global) can occur as a result of changes quite unrelated to Milankovitch cyclicity (Berger 1990). This interval of climatic deterioration, generally considered to have taken place between 11,000 and 10,000 BP, commenced at a time when there was a strong April–June insolation maximum that was unfavourable for ice sheet growth. Yet considerable ice sheet growth took place at this time, possibly as an indirect result of reduced deep water formation in the North Atlantic Ocean (see Chapter 5).

Summary

The principal periods of ice sheet expansion during the Late Quaternary correspond only in part to changes in Milankovitch insolation. A clear relationship appears to have existed during isotope substage 5d when the growth of ice took place at a time when there were not only substantial seasonal changes in insolation but also a relatively deep summer insolation minimum and a strong meridional insolation gradient (Figure 13.2). By contrast, ice accumulation during substage 5b took place during a period of greatly reduced seasonal insolation variations. Glaciation during substage 5b is additionally complicated by the unknown effects of the Atitlan volcanic eruption on global cooling.

The period of rapid northern hemisphere glaciation inferred for the transition between isotope stages 5 and 4 is also associated with marked seasonal changes in insolation. Thus, a summer insolation maximum at 80,000 BP is replaced by an insolation minimum near 75,000 BP that persists until not long after 70,000 BP when it changes to a spring insolation maximum. The mechanisms of glaciation are additionally complicated by the, as yet unknown, effects on global cooling caused by the volcanic eruption at Toba in Sumatra.

During isotope stages 4 and 3, the seasonal signals of Milankovitch insolation are rather weak and are similar to present. Moreover, meridional insolation contrasts for this time interval are scarcely evident. Thus, there is little reason to attribute periods of ice expansion during this time interval to Milankovitch changes. It may be more profitable to envisage the patterns of ice growth and intervening periods of interstadial warmth as fluctuations superimposed upon a pattern of global climate that had already switched to a glacial mode (Berger 1979). The growth of ice sheets in the northern hemisphere during isotope stage 2 was also associated with solar insolation variations that were not as extreme as during earlier time intervals during the Late Quaternary. However, there are grounds to suggest that relatively strong meridional insolation gradients at this time may have been important in promoting widespread glaciation.

DISCUSSION

The evidence that has been provided for Late Quaternary climate change shows that ice sheet glaciation and deglaciation in the northern hemisphere most probably took place on several occasions and that these changes were comparatively rapid. The oxygen isotope stratigraphy from both ocean and ice cores illustrates clearly the rapidity, complexity and high magnitude of some of these changes. However, since oxygen isotope stratigraphy provides a measure of global ice volume changes over time, it cannot provide information about where individual ice masses built up and decayed. Thus, it is at present impossible to estimate the Late Quaternary history of Antarctic ice sheet growth and decay – a factor of crucial importance to any study of changes in the Late Quaternary global climate.

There seems to be a striking correspondence between periods of inferred non-glacial conditions and time intervals when there was a well-defined summer insolation maximum and spring insolation minimum (Figures 13.4 and 13.6). This is very evident for both the last and present interglacials while a similar (although less clearly defined) relationship may exist for several of the intervals of interstadial warmth. A puzzling exception is the interstadial centred on 30,000 BP. By contrast, the periods of ice expansion exhibit a different relationship to Milanko-

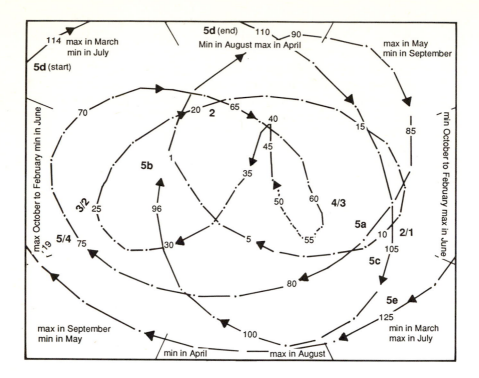

Figure 13.6 Principal components analysis of midmonth daily insolation over the last 130,000 years. Horizontal and vertical axes respectively represent the solstice and equinox axes. The corners of the diagram represent areas where there are particular seasonal insolation minima and maxima. The principal oxygen isotope stages and substages are plotted along the spiral line. Note how all of the 'cold' isotope stage intervals are located in the left of the diagram while most of the 'warm' intervals are located on the right. The only exception to this trend is isotope stage 3. Note also that changes in the insolation signature are relatively rapid between 130,000 and circa 65,000 BP. After circa 65,000 BP the changes are relatively 'sluggish' (see also Figure 13.3) (adapted from Berger 1979).

vitch insolation variations. In general the seasonal rate of change of received insolation appears to have been comparatively rapid between approximately 130,000 BP and 65,000 BP and relatively slow between 65,000 BP and 30,000 BP. As a result, the beginning and end of substage 5d are characterised by a marked difference in the seasonal pattern of insolation (Figure 13.6). By contrast, climatic conditions between 65,000 BP and 30,000 BP appear to have been characterised by a relative absence of strong contrasts in seasonal insolation. Nevertheless, during substage 5d, there appears to be a direct link between diminished summer insolation and ice sheet growth. In other instances (e.g. stage 2), enhanced meridional atmospheric circulation may have been a key factor in causing increased glaciation. By contrast, during the isotope stage 5/4 transition and also possibly during substage 5b there may be a link between glaciation and volcanism.

Caution should be exercised, however, since it may have been the case that the northern hemisphere was not always the leading hemisphere that induced global glaciation (Berger 1979). For example, on several occasions, the strongest summer insolation minimum occurred in the southern hemisphere rather than in the northern hemisphere (Figure 13.3). In this regard, attention should also be given to the suggestion of Broecker and Denton (1990a,b) that the global lowering of temperatures and approximately synchronous expansion of mountain glaciers in both hemispheres during the last glacial maximum cannot be explained solely by Milankovitch mechanisms. They argue that periods of glaciation are approximately synchronous between both hemispheres and that the causes for such changes, as well as changes during glacial terminations, also include fluctuations in atmospheric CO_2 as well as variations in the delivery of fresh water to the oceans and changes in ocean circulation and deep water formation.

Consideration of the elements of the Milankovitch cycles for the Late Quaternary with the principal periods of ice sheet expansion shows that there is some agreement between periods of ice growth (during isotope substages 5d, 5b and stages 4 and 2), and times when there was an insolation decline due to precession of the equinoxes. Additionally, the period of ice sheet expansion known to have taken place in Scandinavia during isotope stage 3 also corresponds with a period of precessional insolation decline (Figure 13.2). Furthermore, the principal periods of ice sheet growth also coincided with times when there were above average amounts of solar radiation received in the low latitudes of the northern hemisphere (Figure 13.2). This has classically been interpreted as indicative of the strong influence of low latitude insolation patterns on ice sheet growth since a strong meridional insolation gradient is necessary to transport moisture to higher latitudes and nourish ice sheet growth (Chappell 1974b). It should be remembered, however, that there may have been considerable lag times between particular Milankovitch insolation changes and the responses of ice sheets and oceans (Ruddiman et al. 1980; Ruddiman and McIntyre 1981c). Furthermore, the response of global climate to a given Milankovitch anomaly also depends on the initial climatic conditions; for example, the extent of sea ice cover, regional albedo, the initial sea level and ocean circulation patterns, etc.

Consideration of the geological evidence and the oxygen isotope stratigraphy suggests that the response of the Earth to changes in Milankovitch insolation variations was highly complex (Figures 13.2 and 13.6). Thus, it may be inappropriate to interpret the observed patterns of Late Quaternary environmental change solely in terms of 40,000 or 20,000 year Milankovitch cycles since some of the geological data cannot be explained in this way. An important step forward for scientists in the future would be to link patterns of Milankovitch insolation variations to a time-dependent advanced general circulation model of the

Earth's atmosphere and oceans. Such a GCM could then be tested against the 'known' patterns of global climate change based on geological evidence. We will have to be patient with this hope, however, since it will be some time before there is a computer sufficiently powerful to accomplish such a task. In the meantime, the main priority for Quaternary scientists should be to continue the thankless task of gathering field data so that existing climate models can be further tested and improved.

REFERENCES

Adamson, D. and Williams, F. 1980 'Structural geology, tectonics and the control of drainage in the Nile Basin'. In M.A.J. Williams and H. Faure (eds) *The Sahara and the Nile*, 225–252, Balkema, Rotterdam.

Alam, M. and Piper, D.J.W. 1977 'Pre-Wisconsin stratigraphy and palaeoclimates off Atlantic Canada and its bearing on glaciation in Quebec'. *Geographie Physique et Quaternaire*, 31, 15–22.

Andersen, B.G. 1979 'The deglaciation of Norway 15,000–10,000 years BP'. *Boreas*, 8, 79–87.

Andersen, B.G. 1981 'Late Weichselian Ice Sheets in Eurasia and Greenland'. In G.H. Denton and T.J. Hughes (eds) *The Last Great Ice Sheets*, 3–66, John Wiley and Sons, New York.

Andersen, B.G., Nydal, R., Wangen, O.P. and S.R. Ostmo 198 'Weichselian before 15,000 years BP at Jaeren-Karmøy in southwestern Norway'. *Boreas*, 10, 297–314.

Andersen, B.G., Wangen, O.P. and S.R. Ostmo 1987 'Quaternary geology of Jaeren and adjacent areas: southwestern Norway'. *Norges Geologiske Undersokelse, Bulletin* 411, 55pp.

Anderson, I. 1987 'Melting glaciers pull the plug of volcanoes?'. *New Scientist*, 12 February, 30.

Andrews, J.T. 1970 'A geomorphological study of post-glacial uplift with particular reference to Arctic Canada'. *Institute of British Geographers Special Publication Number* 2, 156pp.

Andrews, J.T. 1973 'The Wisconsin Laurentide Ice Sheet: Dispersal centers, problems of rates of retreat, and climatic implications'. *Arctic and Alpine Research*, 5, 185–199.

Andrews, J.T. 1982 'On the reconstruction of Pleistocene ice sheets: A review'. *Quaternary Science Reviews*, 1, 1–31.

Andrews, J.T. and Mahaffy, M.A.W. 1976 'Growth rate of the Laurentide Ice Sheet and sea level lowering (with emphasis on the 115,000 BP sea level low)'. *Quaternary Research*, 6, 167–183.

Andrews, J.T. and Miller, G.H. 1984 'Quaternary glacial and non-glacial correlations for the eastern Canadian Arctic'. In R.J. Fulton (ed.) *Quaternary Stratigraphy of Canada – A Canadian Contribution to IGCP Project 24*, Geological Survey of Canada, Paper 84–10, 101–116.

Andrews, J.T., Shilts, W.W. and G.H. Miller 1983 'Multiple deglaciations of the Hudson Bay Lowland, Canada, since deposition of the Missinaibi (last interglacial?) Formation'. *Quaternary Research*, 19, 18–37.

Andrews, J.T., Aksu, A., Kelly, M., Klassen, R., Miller, G.H., Mode, W.N. and P. Mudie 1985 'Land/ocean correlations during the last interglacial/glacial transition, Baffin Bay, northwestern North Atlantic: A review'. *Quaternary Science Reviews*, 4, 333–355.

257

Andrews, J.T., Miller, G.H., Vincent, J.-S. and W.W. Shilts 1986 'Quaternary glaciations in Arctic Canada'. In V. Sibrava, D.Q. Bowen and G.M. Richmond (eds) *Quaternary Glaciations in the Northern Hemisphere*, 243–250, Pergamon Press, Oxford, 514pp.

Anundsen, K. 1985 'Changes in shore-level and ice-front positions in Late-Weichsel and Holocene, southern Norway'. *Norsk Geografisk Tidsskrift*, 39, 205–225.

Arkhipov, S.A. 1984 'Late Pleistocene glaciation of Western Siberia'. In A.A. Velichko (ed.) *Late Quaternary Environments of the Soviet Union*, 13–20, Longman Group Ltd, London.

Armstrong, J.E. 1981 'Post-Vashon Wisconsin glaciation, Fraser Lowland, British Columbia'. *Geological Survey of Canada Bulletin*, 322, 34pp.

Armstrong, J.E. and Clague, J.J. 1977 'Two major Wisconsin lithostratigraphic units in southwest British Columbia'. *Canadian Journal of Earth Sciences*, 14, 1471–1480.

Baker, V.R. 1983a 'Large-scale palaeohydrology'. In K.J. Gregory (ed.) *Background to Palaeohydrology*, 455–478, John Wiley and Sons, Chichester.

Baker, V.R. 1983b 'Late-Pleistocene fluvial systems'. In S.C. Porter (ed.) *Late-Quaternary Environments of the United States, Volume 1, The Late Pleistocene*, 115–129, University of Minnesota Press, Minneapolis.

Baker, V.R. and Bunker, R.C. 1985 'Cataclysmic late Pleistocene flooding from glacial Lake Missoula: a review'. *Quaternary Science Reviews*, 4, 1–41.

Ballantyne, C.K. 1990 'The Late Quaternary glacial history of the Trotternish Escarpment, Isle of Skye, Scotland and its implications for ice-sheet reconstruction'. *Proceedings of the Geologists' Association*, 101, 171–186.

Balling, N. 1980 'The land uplift in Fennoscandia, gravity field anomalies and isostasy'. In N.-A. Mörner (ed.) *Earth Rheology, Isostasy and Eustasy*, 297–322, John Wiley and Sons, Chichester.

Barry, R.G. 1983 'Late-Pleistocene Climatology'. In S.C. Porter (ed.) *Late Quaternary Environments of the United States, Volume 1, The Late Pleistocene*, 390–407, University of Minnesota Press, Minneapolis.

Baulin, V.V. and Danilova, N.S. 1984 'Dynamics of Late Quaternary permafrost in Siberia'. In A.A. Velichko (ed.) *Late Quaternary Environments of the Soviet Union*, 69–78, Longman Group Ltd, London.

Beget, J. 1982 'Recent volcanic activity at Glacier Peak'. *Science*, 215, 1389–1390.

Beget, J. 1983 'Radiocarbon-dated evidence of worldwide early Holocene climatic change'. *Geology*, 11, 389–393.

Beget, J. 1984 'Tephrochronology of Late Wisconsin Deglaciation and Holocene glacier fluctuations near Glacier Peak, North Cascade Range, Washington'. *Quaternary Research*, 21, 304–316.

Behre, K.-E. 1989 'Biostratigraphy of the last glacial period in Europe'. *Quaternary Science Reviews*, 8, 25–44.

Belanger, P.E. 1982 'Paleo-oceanography in the Norwegian sea during the past 130,000 years: coccolithophorid and foraminiferal data'. *Boreas*, 11, 29–36.

Benson, L.V. 1981 'Paleoclimatic significance of lake level fluctuations in the Lahontan Basin'. *Quaternary Research*, 16, 390–403.

Benxing, Z. 1989 'Controversy regarding the existence of a large ice sheet on the Qinghai-Xizang (Tibetan) plateau during the Quaternary period'. *Quaternary Research*, 32, 121–123.

Berger, A.L..1978a 'Long-term variations of caloric insolation resulting from the Earth's orbital elements'. *Quaternary Research*, 9, 139–167.

Berger, A.L. 1978b 'Long term variations of daily insolation and Quaternary climatic change'. *Journal of Atmospheric Science*, 35, 2362–2367.

Berger, A.L. 1979 'Insolation signatures of Quaternary climatic changes'. *Il Nuovo Cimento*, 2C, 1, 63–87.

Berger, W.H. 1990 'The Younger Dryas cold spell – a quest for causes'. *Paleogeography,*

Paleoclimatology, Paleoecology (Global and Planetary Change Section), 89, 219–237.

Berglund, B.E. and Lagerlund, E. 1981 'Eemian and Weichselian stratigraphy in South Sweden'. *Boreas*, 10, 323–362.

Bigarella, J.J. and Andrade, G.D. de 1965 'Contribution to the study of the Brazilian Quaternary'. In H.E. Wright and D.G. Frey (eds) *International Studies on the Quaternary*, Geological Society of America Special Paper 84, 433–451.

Billard, A. and Orombelli, G. 1986 'Quaternary glaciations in the French and Italian Piedmonts of the Alps'. In V. Sibrava, D.Q. Bowen and G.M. Richmond (eds) *Quaternary Glaciations in the Northern Hemisphere*, 407–412, Pergamon Press, Oxford.

Bjorck, S. and Digerfeldt, G. 1986 'Late Weichselian–Early Holocene shore displacement west of Mt. Billingen, within the Middle Swedish end-moraine zone'. *Boreas*, 15, 1–18.

Black, R.F. 1969 'Climatically significant fossil periglacial phenomena in northcentral United States'. *Biuletyn Peryglacjalny*, 20, 225–238.

Black, R.F. 1980 'Isostatic, tectonic and eustatic movements of sea level in the Aleutian Islands, Alaska'. In N.-A. Mörner (ed.) *Earth Rheology, Isostasy and Eustasy*, 231–250, John Wiley and Sons, Chichester, 599pp.

Blagbrough, J.W. and Farkas, S.E. 1968 'Rock glaciers in San Mateo Mountains, south-central New Mexico'. *American Journal of Science*, 226, 812–823.

Bloom, A.L. 1967 'Pleistocene shorelines: a new test of isostasy'. *Geological Society of America Bulletin*, 78, 1477–1494.

Bloom, A.L. 1971 'Glacial-eustatic and isotatic controls of sea level since the last glaciation'. In K.K. Turekian (ed.) *The Late Cenozoic Glacial Ages*, 355–379, Yale University Press, New Haven, Connecticut.

Bloom, A.L., Broecker, W.S., Chappell, J., Mathews, R.K. and K.J. Mesolella 1974 'Quaternary sea level fluctuations on a tectonic coast: new ^{230}Th/^{234}U dates from the Huon Peninsula, New Guinea'. *Quaternary Research*, 4, 185–205.

Bogaard, P. and Schmincke, H.-U. 1985 'Laacher See Tephra: A widespread isochronous late Quaternary tephra layer in central and northern Europe'. *Geological Society of America Bulletin*, 96, 1554–1571.

Bonatti, E. and Gartner, S. 1973 'North Mediterranean climate during the last Wurm Glaciation', *Nature*, 209, 984.

Boulton, G.S. 1979 'Glacial history of the Spitsbergen archilpelago and the problem of a Barents Shelf ice sheet'. *Boreas*, 8, 31–57.

Boulton, G.S., Jones, A.S., Clayton, K.M. and M.J. Kenning 1977 'A British ice sheet model and patterns of glacial erosion and deposition in Britain'. In F.W. Shotton (ed.) *British Quaternary Studies – Recent Advances*, 231–246, Oxford University Press, Oxford.

Bowen, D.Q. 1978 *Quaternary Geology*. Pergamon, Oxford, 221pp.

Bowler, J.M. 1978 'Glacial age aeolian events at high and low latitudes: a southern hemisphere perspective'. In E.M. Van Zinderen Bakker (ed.) *Antarctic Glacial History and World Palaeoenvironments*, 149–172, Balkema, Rotterdam.

Bowler, J.M., Hope, G.S., Jennings, J.N., Singh, G. and D. Walker, 1976 'Late Quaternary climates of Australia and New Guinea'. *Quaternary Research*, 6, 359–394.

Bowles, F.A., Jack, R.N. and I.S.E. Carmichael 1973 'Investigation of deep-sea volcanic ash layers from equatorial Pacific cores'. *Geological Society of America Bulletin*, 84, 2371–2388.

Bradley, R.S. 1985 *Quaternary Palaeoclimatology*. Allen and Unwin, Boston, 472pp.

Bramlette, M.N. and Bradley, W.H. 1941 'Lithology and geological interpretations: geology and biology of North Atlantic deep-sea cores between Newfoundland and Ireland'. *US Geological Survey Professional Paper* 196A, 1–34.

Bretz, J.H. 1969 'The Lake Missoula floods and the Channeled Scabland'. *Journal of Geology*, 77, 505–543.

Briggs, D.J. and Gilbertson, D.D. 1980 'Quaternary processes and environments in the Upper Thames valley'. *Transactions Institute of British Geographers, New Series*, 5, 53–65.

Broccoli, A.J. and Manabe, S. 1987 'The effects of the Laurentide Ice Sheet on North American climate during the last glacial maximum'. *Geographie Physique et Quaternaire*, 41, 291–300.

Broecker, W.S. and Denton, G.H. 1990a 'What drives glacial cycles?'. *Scientific American*, January 1990, 43–50.

Broecker, W.S. and Denton, G.H. 1990b 'The role of Ocean-Atmosphere Reorganisations in Glacial Cycles'. *Quaternary Science Reviews*, 9, 305–343.

Broecker, W.S. and Orr, P.C. 1958 'Radiocarbon chronology of Lake Lahontan and Lake Bonneville'. *Geological Society of America Bulletin*, 70, 1009–32.

Broecker, W.S., Thurber, D.L., Goddard, J., Ku, T.-L., Mathews, R.K. and K.J. Mesolella, 1968 'Milankovitch hypothesis supported by precise dating of coral reefs and deep sea sediments'. *Science*, 159, 297–300.

Broecker, W.S., Andree, M., Bonani, G., Wolfi, W., Oeschger, H. and M. Klas 1988 'Can the Greenland climatic jumps be identified in records from ocean and land?'. *Quaternary Research*, 30, 1–6.

Broecker, W.S., Kennett, J.P., Teller, J., Trumbore, S., Bonani, G. and W. Wolfli 1989 'Routing of meltwater from the Laurentide Ice Sheet during the Younger Dryas cold episode'. *Nature*, 341, 318–321.

Bull, W.B. and Cooper, A.F. 1986 'Uplifted marine terraces along the Alpine fault, New Zealand'. *Science*, 234, 1225–1228.

Butzer, K.W. 1980 'Pleistocene history of the Nile Valley in Egypt and Lower Nubia'. In M.A.J. Williams and H. Faure (eds) *The Sahara and the Nile*, 253–280, Balkema, Rotterdam.

Butzer, K.W. and Cuerda, J. 1962 'Coastal stratigraphy of southern Mallorca and its implications for the Pleistocene chronology of the Mediterranean Sea'. *Journal of Geology*, 70, 398–416.

Caldenius, C. 1932 'Las glaciaciones cuaternarias en la Patagonia y Tierra del Fuego'. *Geografiska Annaler*, 14, 1–61.

Catt, J.A. 1988 *Quaternary Geology for Scientists and Engineers*, Ellis Horwood Ltd, Chichester, 340pp.

Chappell, J. 1974a 'Geology of coral terraces, Huon Peninsula, New Guinea: a study of Quaternary tectonic movements and sea level changes'. *Geological Society of America Bulletin*, 85, 553–570.

Chappell, J. 1974b 'Late Quaternary glacio- and hydro-isostasy on a layered earth'. *Quaternary Research*, 4, 405–428.

Chappell, J. 1974c 'Relationships between sea levels, ^{18}O variations and orbital perturbations, during the past 250,000 years'. *Nature*, 252, 199–202.

Chappell, J. 1983 'A revised sea-level record for the last 300,000 years from Papua New Guinea'. *Search*, 14, 99–101.

Chappell, J. 1987 'Late Quaternary sea level changes in the Australian region'. In M.J. Tooley and I. Shennan (eds) *Sea Level Changes*, 296–331, Basil Blackwell, Oxford

Chappell, J. and Shackleton, N.J. 1986 'Oxygen isotopes and sea-level'. *Nature*, 324, 137–140.

Chesner, C.A., Rose, W.I., Deino, A., Drake, R. and J. Westgate 1990 'Eruptive history of Earth's largest Quaternary caldera (Toba, Indonesia) clarified'. *Geology*, 19, 200–203.

Chorley, R.J., Schumm, S.A. and D.E. Sugden 1984 *Geomorphology*, Methuen Press, London.

Clague, J.J. 1975 'Glacier flow patterns and the origin of the Late Wisconsinan till in the southern Rocky Mountain trench, British Columbia'. *Geological Society of America Bulletin*, 86, 721–731.

Clague, J.J. 1981 'Late Quaternary geology and geochronology of British Columbia, Part 2: Summary and discussion of radiocarbon-dated Quaternary history'. *Geological Survey of Canada*, Paper 80–35, 41pp.

Clague, J.J. and Mathews, W.H. 1973 'The magnitude of jökulhlaups'. *Journal of Glaciology*, 12, 501–504.

Clague, J.J., Armstrong, R.W. and W.H. Mathews 1982 'Late Quaternary geology of eastern Graham Island, Queen Charlotte islands, British Columbia'. *Canadian Journal of Earth Sciences*, 19, 1786–1795.

Clapperton, C.M. 1983 'The Glaciation of the Andes'. *Quaternary Science Reviews*, 2, 83–155.

Clapperton, C.M. 1985 'Significance of a late-glacial readvance in the Ecuadorian Andes'. In J. Rabassa (ed.) *Quaternary of South America and Antarctic Peninsula*, 149–158, Balkema, Rotterdam.

Clapperton, C.M. 1987 'Glacial geomorphology, Quaternary glacial sequence and palaeo-climatic inferences in the Ecuadorian Andes'. In V. Gardiner (ed.) *International Geomorphology*, Part II, 843–870, John Wiley and Sons, Chichester.

Clapperton, C.M. and McEwan, C. 1985 'Late Quaternary moraines in the Chimborazo area, Ecuador'. *Arctic and Alpine Research*, 17, 135–142.

Clapperton, C.M. and Roberts, D.E. 1986 'Quaternary sea level changes in the Falkland Islands'. In J. Rabassa (ed.) *Quaternary of South America and Antarctic Peninsula*, 4, 99–117, Balkema, Rotterdam.

Clapperton, C.M. and Sugden, D.E. 1990 'Late Cenozoic Glacial History of the Ross Embayment, Antarctica'. *Quaternary Science Reviews*, 9, 253–272.

Clark, J.A. 1980 'A numerical model of worldwide sea-level changes on a viscoelastic Earth'. In N.-A. Mörner (ed.) *Earth Rheology, Isostasy and Eustasy*, 525–534, John Wiley and Sons, Chichester, 599pp.

Clark, J.A., Farrell, W.E. and W.R. Peltier 1978 'Global changes in postglacial sea level: a numerical calculation'. *Quaternary Research*, 9, 265–287.

Clayton, L. 1983 'Chronology of Lake Agassiz drainage to Lake Superior'. In J.T. Teller and L. Clayton (eds) *Glacial Lake Agassiz*, 291–308, Geological Association of Canada Special Paper 26.

Clayton, L. and Moran, S.R. 1982 Chronology of Late Wisconsinan glaciation in middle North America. *Quaternary Science Reviews*, 1, 55–82.

Clayton, L., Teller, J.T. and J.W. Attig 1985 'Surging of the southwestern part of the Laurentide Ice Sheet'. *Boreas*, 14, 235–241.

CLIMAP Project Members 1976 'The surface of the Ice Age Earth'. *Science*, 191, 1131–1136.

Coope, G.R. and Pennington, W. 1977 'The Windermere Interstadial of the Late Devensian'. *Philosophical Transactions of the Royal Society of London*, B280, 337–339.

Craig, H. 1965 'The measurement of oxygen isotope paleotemperature'. In *Proceedings of the Spoleto Conference of Stable Isotopes in Oceanographic Studies and Paleotemperatures*, 3–24, Consiglio Nazionale delle Richerche Laboratoriodi Geologia Nucleare, Pisa.

Cramp, A., Vitaliano, C.J. and M.B. Collins 1989 'Identification and dispersion of the Campanian ash layer (Y-5) in the sediments of the Eastern Mediterranean'. *Geo-Marine Letters*, 9, 19–25.

Crittenden, M.D. Jr 1963 'Effective viscosity of the Earth derived from isostatic loading of Pleistocene Lake Bonneville'. *Journal of Geophysical Research*, 68, 5517–30.

Croll, J. 1867a 'On the eccentricity of the Earth's orbit, and its physical relations to the glacial epoch'. *Philosophical Magazine*, 33, 119–131.

Croll, J. 1867b 'On the change in the obliquity of the ecliptic, its influence on the climate of the polar regions and on the level of the sea'. *Philosophical Magazine*, 33, 426–445.

Croll, J. 1875 *Climate and Time*, Appleton and Co, New York.

Currey, D.R. 1980 'Coastal geomorphology of Great Salt Lake and vicinity'. In J.W. Gwynn (ed.) *Great Salt Lake: A scientific, historic and economic overview*, 69–82, Utah Geological and Mineral Survey Bulletin 116.

Damuth, J.E. and Fairbridge, R.W. 1970 'Arkosic sands of the last glacial stage in the tropical Atlantic off Brazil'. *Geological Society of America Bulletin*, 81, 189–206.

Dansgaard, W.S. 1984 'Selected climates from the past and their relevance to possible future climate'. In H. Flohn and R. Fantechi (eds) *The Climate of Europe: Past, Present and Future*, 208–213, D Reidel, Dordrecht, 356pp.

Dansgaard, W.S. and Tauber, H. 1969 'Glacier oxygen 18 content and Pleistocene ocean temperatures'. *Science*, 166, 499–502.

Dansgaard, W., Johnsen, S.J., Clausen, H.B. and C.C. Langway Jr 1971 'Climatic record revealed by Camp Century Ice Core'. In K.K. Turekian (ed.) *The Late Cenozoic Glacial Ages*, 37–56, Yale University Press, New Haven, Connecticut, 606pp.

Dansgaard, W.S., Clausen, H.B., Gundestrup, N., Hammer, C.U., Johnsen, S.F., Kristin-dottir, P.M. and N. Reeh 1982 'A new Greenland deep ice core'. *Science*, 218, 1273–1277.

Dansgaard, W., White, J.W.C. and S.J. Johnson 1989 'The abrupt termination of the Younger Dryas climatic event'. *Nature*, 339, 532–533.

Dawson, A.G. 1984 'Quaternary Sea-Level Changes in Western Scotland'. *Quaternary Science Reviews*, 3, 345–368.

De Beaulieu, J.L. and Reille, M. 1984 'A long Upper Pleistocene pollen record from Les Echets, near Lyon, France'. *Boreas*, 13, 111–132.

Denton, G.H. and Hughes, T.J. 1981 *The Last Great Ice Sheets*, John Wiley and Sons, New York, 484pp.

Denton, G.H., Armstrong, R.L. and M. Stuiver 1971 'The Late Cenozoic glacial history of Antarctica'. In K. Turekian (ed.) *The Late Cenozoic Glacial Ages*, 267–306, Yale University Press, New Haven, Connecticut.

Diester-Haas, L. 1976 'Late Quaternary climatic variation in northwest Africa deduced from East Atlantic sediment cores'. *Quaternary Research*, 6, 299–314.

Diester-Haas, L. 1980 'Upwelling and climate off Northwest Africa during the Late Quaternary'. *Palaeoecology of Africa*, 12, 229–238.

Dodge, R.E., Fairbanks, R.G., Benninger, L.K. and F. Maurrasse 1983 'Pleistocene sea levels from raised coral reefs of Haiti'. *Science*, 219, 1423–1425.

Donner, J.J. 1980 'The determination and dating of synchronous Late Quaternary shore-lines in Fennoscandia'. In N.A. Mörner (ed.) *Earth Rheology, Isostasy and Eustasy*, 285–293, John Wiley and Sons, New York.

Donner, J.J. and Eronen, M. 1981 'Stages of the Baltic Sea and Late Quaternary Shoreline Displacement in Finland'. *INQUA Subcommision on the Shorelines of Northwestern Europe Excursion Guide 9–14 September 1981, Stencil No. 5*, University of Helsinki, 53pp.

Dredge, L.A. and Thorleifson, H. 1987 'The Middle Wisconsin History of the Laurentide Ice Sheet'. *Geographie Physique et Quaternaire*, 41, 215–235.

Dreimanis, A. and Raukas, A. 1975 'Did Middle Wisconsin, Middle Weichselian, and their equivalents represent an interglacial or an interstadial complex in the Northern Hemisphere?'. In R.P. Suggate and M.M. Cressell (eds) *Quaternary Studies*, 109–120, The Royal Society of New Zealand, Wellington.

Dreimanis, A., Terasmae, J and G.D. McKenzie 1966 'The Port Talbot interstade of the Wisconsin glaciation'. *Canadian Journal of Earth Sciences*, 3, 305–325.

Drewry, D.J. 1983 *Antarctica: Glaciological and Geophysical Folio*, Scott Polar Research Institute, University of Cambridge.

Drexler, J.W., Rose, W.I. Jr, Sparks, R.S.J. and M.T. Ledbetter 1980 'The Los Chocoyos Ash, Guatemala: A major stratigraphic marker in Middle America and in Three Ocean Basins'. *Quaternary Research*, 13, 327–345.

Dubois, J.-M. and Dionne, J.-C. 1985 'The Quebec North Shore Moraine system: a major feature of Late Wisconsin deglaciation'. In H.W. Borns, P. LaSalle and W.B. Thompson (eds) *Late Pleistocene History of Northeastern New England and Adjacent Quebec*, Geological Society of America Special Paper 197, 125–133.

Dugmore, A. 1989 'Icelandic Volcanic Ash in Scotland'. *Scottish Geographical Magazine*, 105, 3, 168–172.

Duplessy, J.C., Chenouard, L. and F. Vila 1975 'Weyl's theory of glaciation supported by isotopic study of Norwegian core K 11'. *Science*, 188, 1208–1209.

Duplessy, J.C., Moyes, J. and C. Pujol 1980 'Deep water formation in the North Atlantic Ocean during the last ice age'. *Nature*, 290, 479–482.

Dyke, A.S. 1983 'Quaternary Geology of Somerset Island, District of Franklin'. *Geological Survey of Canada, Memoir* 404, 32pp.

Dyke, A.S. and Prest, V.K. 1987 'Late Wisconsinan and Holocene History of the Laurentide Ice Sheet'. *Geographie Physique et Quaternaire*, 41, 237–264.

Easterbrook, D.J. 1976 'Quaternary geology of the Pacific Northwest'. In W.C. Mahaney (ed.) *Quaternary Stratigraphy of North America*, 441–462, Dowden, Hutchinson and Ross, Stroudsburg, Pennsylvania.

Emiliani, C. 1955 'Pleistocene temperatures'. *Journal of Geology*, 63, 538–575.

Emiliani, C. 1961 'Cenozoic climatic changes as indicated by the stratigraphy and chronology of deep-sea cores of Globigerina facies'. *Annals of the New York Academy of Science*, 95, 521–536.

Emiliani, C. 1966 'Isotopic palaeotemperatures'. *Science*, 154, 851–857.

Ericson, D.B., Broecker, W.S., Kulp, J.L. and G. Wollin 1956 'Late Pleistocene climate and deep-sea sediments'. *Science*, 124, 385–389.

Eronen, M. 1974 'The history of the Litorina Sea and associated Holocene events'. *Societas Scientiarum Fennica Helsinki-Helsingfors, Commentationes Physico-Mathematicae*, 44, 79–195.

Eronen, M. 1983 'Late Weichselian and Holocene shore displacement in Finland'. In D.E. Smith and A.G. Dawson (eds) *Shorelines and Isostasy*, 183–208, Academic Press, London.

Eronen, M. and Olander, H. 1990 'On the world's ice ages and changing environments', *Report* YJT-90-13, Nuclear Waste Commission of Finnish Power Companies, 72pp, Helsinki.

Evenson, E.B. and Dreimanis, A. 1976 'Late glacial (14,000–10,000 years BP) history of the Great Lakes region and possible correlations'. In *Quaternary Glaciations in the Northern Hemisphere*, Report 3, International Geological Correlation Program Project 73/1/24, 217–238.

Ewing, M. and Donn, W.L. 1956 'A theory of ice ages'. *Science*, 123, 1061–1066.

Ewing, M. and Donn, W.L. 1958 'A theory of ice ages II'. *Science*, 127, 1159–1162.

Ewing, M., Ericson, D.B. and B.C. Heezen 1958 'Sediments and topography of the Gulf of Mexico'. In L.G. Weeks (ed.) *Habitat of Oil*, American Association of Petroleum Geologists, 995–1053, Tulsa, Oklahoma.

Eyles, N. and McCabe, A.M. 1989 'The Late Devensian (< 20,000 BP) Irish Sea Basin: The Sedimentary Record of a Collapsed Ice Sheet Margin'. *Quaternary Science Reviews*, 8, 307–353.

Fairbanks, R.G. 1989 'A 17,000-year glacio-eustatic sea level record: influence of glacial melting rates on the Younger Dryas event and deep-ocean circulation'. *Nature*, 342, 637–642.

Fairbridge, R.W. 1983 'Isostasy and Eustasy'. In D.E. Smith and A.G. Dawson (eds) *Shorelines and Isostasy*, 3–28, Academic Press, London.

Farrand, W.R. 1971 'Late Quaternary paleoclimates of the eastern Mediterranean area'. In K.K. Turekian (ed.) *The Late Cenozoic Glacial Ages*, 529–64, Yale University Press, New Haven, Connecticut.

Fisher, D.A., Reeh, N. and K. Langley 1985 'Objective reconstructions of the Late Wisconsinan Laurentide ice sheet and the significance of deformable beds'. *Geographie Physique et Quaternaire*, 39, 229–238.

Fisher, R.V. and Schmincke, H.-U. 1984 *Pyroclastic Rocks*. Springer-Verlag, Berlin, 472pp.

Fjeldskaar, W. and Kanestrom, R. 1980 'Younger Dryas geoid-deformation caused by deglaciation in Fennoscandia'. In N.-A. Mörner (ed.) *Earth Rheology, Isostasy and Eustasy*, 569–576, John Wiley and Sons, Chichester, 599pp.

Flemal, R.C. 1976 'Pingos and pingo scars: Their characteristics, distribution and utility in reconstructing former permafrost environments'. *Quaternary Research*, 6, 37–53.

Flenley, J.R. 1979 *The Equatorial Rain Forest: A Geological History*, Butterworths, London and Boston, 162pp.

Flint, R.F. 1943 'Growth of the North American ice sheet during the Wisconsin Age'. *Geological Society of America Bulletin*, 54, 325–362.

Flint, R.F. 1971 *Glacial and Quaternary Geology*, John Wiley and Sons, New York, 892pp.

French, H.M. 1976 *The Periglacial Environment*, Longman Group Ltd, London, 309pp.

Frenzel, B. 1973 *Climatic Fluctuations of the Ice Age*. The Press of Case Western Reserve University, Cleveland and London, 306pp.

Friedman, I. 1977 'The Amazon Basin, another Sahel?'. *Science*, 197, 7.

Fullerton, D.S. 1980 'Preliminary Correlation of Post-Erie Interstadial Events (16,000–10,000 Radiocarbon Years before Present), Central and Eastern Great Lakes Region, and Hudson, Champlain, and St. Lawrence Lowlands, United States and Canada'. *United States Geological Survey Professional Paper* 1089.

Fullerton, D.S. and Richmond, G.M. 1986 'Comparison of the marine oxygen isotope record, the eustatic sea level record, and the chronology of glaciation in the United States of America'. In V. Sibrava, D.Q. Bowen and G.M. Richmond (eds) *Quaternary Glaciations in the Northern Hemisphere*, 197–200, Pergamon Press, Oxford, 514pp.

Fulton, R.J. 1986 'Quaternary stratigraphy of Canada'. In V. Sibrava, D.Q. Bowen and G.M. Richmond (eds) *Quaternary Glaciations in the Northern Hemisphere*, 207–210, Pergamon Press, Oxford, 514pp.

Fulton, R.J. (ed.) 1989 *Quaternary Geology of Canada and Greenland. Geological Survey of Canada, no. 1* (Geological Society of America, The Geology of North America v. K-1), 839pp.

Fulton, R.J. and Prest, V.K. 1987 'Introduction: The Laurentide Ice Sheet and its significance'. *Geographie Physique et Quaternaire*, 41, 181–186.

Fulton, R.J., Fenton, M.M. and N.W. Rutter 1986a 'Summary of Quaternary stratigraphy and history, Western Canada'. In V. Sibrava, D.Q. Bowen and G.M. Richmond (eds) *Quaternary Glaciations in the Northern Hemisphere*, 229–242, Pergamon Press, Oxford, 514pp.

Fulton, R.J., Karrow, P.F., Lasalle, P. and D.R. Grant 1986b 'Summary of Quaternary stratigraphy and history, eastern Canada'. In V. Sibrava, D.Q. Bowen and G.M. Richmond (eds) *Quaternary Glaciations in the Northern Hemisphere*, 211–228, Pergamon, Oxford, 514pp.

Funder, S. 1989 'Quaternary Geology of the ice-free areas and adjacent shelves of Greenland'. In R.J. Fulton (ed.) *Quaternary Geology of Canada and Greenland. Geological Survey of Canada, Geology of Canada, no. 1* (Geological Society of America, The Geology of North America v. K-1), 743–792.

Funder, S. and Hjort, C. 1973 'Aspects of the Weichselian chronology in central East Greenland'. *Boreas*, 2, 69–84.

Galloway, R.W. 1970 'The full-glacial climate of southwestern United States'. *Annals of the Association of American Geographers*, 60, 245–256.

Gaposchkin, E.M. 1973 'Satellite dynamics'. In E.M. Gaposchkin (ed.) *Smithsonian *standard earth III*, Smithsonian Astronomical Observatory Special Report, 353, 85–192.

Gascoyne, M., Schwarcz, H.P. and D.C. Ford. 1980 'A palaeotemperature record for the Mid-Wisconsin in Vancouver Island'. *Nature*, 285, 474–476.

Gasse, F. 1977 'Evolution of Lake Abhe (Ethiopia and T.F.A.I.) from 70,000 BP'. *Nature*, 265, 42–45.

Gasse, F., Rognon, P. and F.A. Street 1980 'Quaternary history of the Afar and Ethiopian Rift Lakes'. In M.A.J. Williams and H. Faure (eds) *The Sahara and the Nile*, 361–400, Balkema, Rotterdam.

Gates, W.L. 1976a 'The numerical simulation of Ice-Age climate with a global general circulation model'. *Journal of Atmospheric Sciences*, 33, 1844–1873.

Gates, W.L. 1976b 'Modelling the Ice Age climate'. *Science*, 191, 1138–1144.

Gilbert, G.K. 1890 *Lake Bonneville*, United States Geological Survey Monograph 1.

Goosens, D. 1988 'Scale model simulations of the deposition of loess in hilly terrain'. *Earth Surface Processes and Landforms*, 13, 533–544.

Goudie, A.S. 1983 *Environmental Change*, Clarendon Press, Oxford, 258pp.

Goudie, A.S., Allchin, B. and K.T.M. Hedge 1973 'The former extensions of the Great Indian Sand Desert'. *Geographical Journal*, 139, 243–257.

Gow, A.J. and Williamson, T. 1971 'Volcanic ash in the Antarctic Ice Sheet and its possible climatic implications'. *Earth and Planetary Science Letters*, 13, 210–218.

Grant, D.R. 1987 'Glacial style and ice limits, the Quaternary stratigraphic record, and changes of land and ocean level in the Atlantic Provinces, Canada'. *Geographie Physique et Quaternaire*, 31, 347–360.

Grant, D.R. 1989 'Quaternary geology of the Atlantic Appalachian region of Canada'. In R.J. Fulton (ed.) *Quaternary Geology of Canada and Greenland*, 393–440.

Grant, D.R. and King, L.H. 1984 'A stratigraphic framework for the Quaternary history of the Atlantic Provinces, Canada'. In R.J. Fulton (ed.) *Quaternary Stratigraphy of Canada – A Canadian Contribution to IGCP Project 24*, Geological Survey of Canada, Paper 84–10, 173–191.

Gray, J.M. 1974 'The Main Rock Platform of the Firth of Lorn, Western Scotland'. *Transactions of the Institute of British Geographers*, 61, 81–99.

Gray, J.M. 1978 'Low-level shore platforms in the south-west Scottish Highlands: altitude, age and correlation'. *Transactions of the Institute of British Geographers, New Series*, 3, 151–164.

Green, C.P. and MacGregor, D.F.M. 1980 'Quaternary evolution of the River Thames'. In D.K.C. Jones (ed.) *The Shaping of Southern England*, Institute of British Geographers Special Publication No. 11, 177–202.

Grosswald, M.G. 1980 'Late Weichselian Ice Sheet of Northern Eurasia'. *Quaternary Research*, 13, 1–32.

Grosswald, M.G. 1984 'Glaciation of the continental shelves (Parts I and II)'. *Polar Geography and Geology*, 8, 3, 194–258 and 287–351.

Grove, A.T. 1969 'Landforms and climatic change in the Kalahari and Ngamiland'. *Geographical Journal*, 135, 192–212.

Grove, A.T. and Warren, A. 1968 'Quaternary landforms and climate on the south side of the Sahara'. *Geographical Journal*, 134, 194–208.

Haesaerts, P. 1984 'Aspects de l'evolution du paysage et de l'environment en Belgique au Quaternaire'. Chapitre III; *Peuples chasseurs de la Belgique prehistorique dans leur cadre naturel*, Publication de L'Institute Royale Science Naturel de Belgique, 27–40.

Hafsten, U. 1983 'Biostratigraphical evidence for Late Weichselian and Holocene sea level changes in southern Norway'. In D.E. Smith and A.G. Dawson (eds) *Shorelines and Isostasy*, 161–182, Academic Press, London, 387pp.

Hahn, G.A., Rose, W.I. Jr and T. Meyers 1979 'Geochemical correlation of genetically related rhyolitic ash-flow and air-fall ashes, central and western Guatemala and the equatorial Pacific'. In W. Elston and C. Chapin (eds) *Ash Flow Tuffs*, Geological Society of America Special Paper 180, 101–112.

Hall, K. 1982 'Rapid deglaciation as an initiator of volcanic activity: an hypothesis'. *Earth Surface Processes and Landforms*, 7, 45–51.

Hallberg, G.R. 1986 'Pre-Wisconsin glacial stratigraphy of the Central Plains Region in Iowa, Nebraska, Kansas and Missouri'. In V. Sibrava, D.Q. Bowen and G.M. Richmond (eds) *Quaternary Glaciations in the Northern Hemisphere*, 11–16, Pergamon Press, Oxford, 514pp.

Hamilton, A.C. 1976 'The significance of patterns of distribution shown by forest plants and animals in tropical Africa for the reconstruction of upper Pleistocene palaeo-environments: a review'. *Palaeoecology of Africa*, 9, 63–97.

Hamilton, A.C. 1982 *Environmental History of East Africa: a study of the Quaternary*, Academic Press, London, 328pp.

Hamilton, T. 1986a 'Correlation of Quaternary Glacial deposits in Alaska'. In V. Sibrava, D.Q. Bowen and G.M. Richmond (eds) *Quaternary Glaciations in the Northern Hemisphere*, 171–180, Pergamon Press, Oxford, 514pp.

Hamilton, T.D. 1986b 'Late Cenozoic Glaciation of the Central Brooks Range'. In T.D. Hamilton, K.M. Reed and R.M. Thorson (eds) *Glaciations in Alaska*, Alaska Geological Society, 9–49.

Hamilton, T.D. and Thorson, R.M. 1983 'The Cordilleran Ice Sheet in Alaska'. In S.C. Porter (ed.) *Late-Quaternary Environments of the United States, Volume 1, The Late Pleistocene*, 38–52, University of Minnesota Press, Minneapolis.

Hammer, C.U., Clausen, H.B. and W. Dansgaard 1981 'Past volcansim and climate revealed by Greenland ice cores'. *Journal of Volcanology and Geothermal Research*, 11, 3–10.

Hays, J.D., Lozano, J.A., Shackleton, N. and G. Irving 1976 'Reconstruction of the Atlantic and western Indian sectors of the 18,000 B.P. Antarctic Ocean'. In R.M. Cline and J.D. Hays (eds) *Investigation of Late Quaternary Paleoceanography and Paleoclimatology*, Geological Society of America Memoir 145, 337–374.

Heine, K. 1982 'The main stages of the late Quaternary evolution of the Kalahari region, southern Africa'. *Palaeoecology of Africa*, 15, 53–76.

Herail, G., Huschman, J. and G. Jalut 1986 'Quaternary glaciations in the French Pyrenees'. In V. Sibrava, D.Q. Bowen and G.M. Richmond (eds) *Quaternary Glaciations in the Northern Hemisphere*, 397–402, Pergamon Press, Oxford, 514pp.

Heusser, C.J. 1989 'Climate and chronology of Antarctica and adjacent South America over the past 30,000 yr'. *Palaeogeography, Palaeoclimatology, Palaeoecology*, 76, 31–37.

Hey, R.W. 1978 'Horizontal Quaternary shorelines of the Mediterranean'. *Quaternary Research*, 10, 197–203.

Hill, P.R., Mudie, P.J., Moran, K. and S.M. Blasco 1985 'A sea level curve for the Canadian Beaufort Shelf'. *Canadian Journal of Earth Sciences*, 22, 1383–1393.

Hillaire-Marcel, C., Ochietti, S. and J.-S. Vincent 1981 'Saskami moraine, Quebec: a 500 km long moraine without climatic control'. *Geology*, 9, 210–214.

Hirvas, H., Korpela, K. and R. Kujansuu 1981 'Weichselian in Finland before 15,000 B.P'. *Boreas*, 10, 423–431.

Hodgson, D.A. and Vincent, J.-S. 1984 'A 10,000 yr. BP extensive ice shelf over Viscount Melville Sound, Arctic Canada'. *Quaternary Research*, 22, 18–30.

Hollin, J.T. 1962 'On the glacial history of Antarctica'. *Journal of Glaciology*, 4, 173–195.

Hollin, J.T. 1969 'Ice-sheet surges and the geological record'. *Canadian Journal of Earth Sciences*, 5, 903–910.

Holmes, R. 1977 'Quaternary deposits of the central North Sea, 5. The Quaternary geology of the UK sector of the North sea between 56 and 58°N'. *Report of the Insti-*

tute of Geological Sciences, No. 77/14.

Hooke, R. LeB. 1972 'Geomorphic evidence for Late Wisconsin and Holocene tectonic deformation, Death Valley, California'. *Geological Society of America Bulletin*, 83, 2073–98.

Hope, G.S., Peterson, J.A., Radok, U. and I. Allison 1976 *The Equatorial Glaciers of New Guinea*, Balkema, Rotterdam, 244pp.

Hopkins, D.M. 1972 'The palaeogeography and climatic history of Beringia during late Cenozoic time'. *Inter-Nord* 12, 121–150.

Hopkins, D.M. 1973 'Sea level history in Beringia during the past 250,000 years'. *Quaternary Research*, 3, 520–540.

Hopley, D. 1983 'Deformation of the North Queensland continental shelf in the Late Quaternary'. In D.E. Smith and A.G. Dawson (eds) *Shorelines and Isostasy*, 347–368, Academic Press, London.

Houmark-Nielsen, M. and Kolstrup, E. 1981 'A radiocarbon-dated Weichselian sequence from Sejero, Denmark'. *Geologiska Foreningen i Stockholm, Forhandlingen*, 103, 73–78.

Huairen, Y. and Xi-qing, C. 1987 'Quaternary transgressions, eustatic changes and movements of shorelines in North and East China'. In V. Gardiner (ed.) *International Geomorphology*, Part II, 807–827, John Wiley and Sons, Chichester.

Huairen, Y., Xi-qing, C. and X. Zhiren 1986 'Sea-level changes since the last deglaciation and its impact on the East China lowlands'. In Q. Yunshan and Z. Songling (eds) *Late Quaternary Sea-level Changes*, 199–212, China Ocean Press, Beijing.

Hughes, T.J., Denton, G.H. and M.G. Grosswald 1977 'Was there a late Wurm Arctic Ice Sheet?' *Nature*, 266, 596–602.

Hughes, T.J., Denton, G.H., Andersen, B.G., Schilling, D.H., Fastook, J.L. and C.S. Lingle 1981 'The Last Great Ice Sheets: a global view'. In G.H. Denton and T.J. Hughes (eds) *The Last Great Ice Sheets*, John Wiley and Sons, 275–318.

Hyvarinen, H. 1973 'The deglaciation history of eastern Fennoscandia – recent data from Finland'. *Boreas*, 2, 85–102.

Imbrie, J. and Imbrie, K.P. 1979 *Ice Ages: solving the mystery*, Macmillan, London, 229pp.

Jansen, E., Befring, S., Bugge, T., Eidvin, T., Holtedahl, H. and H.-P. Sejrup 1987 'Large submarine slides on the Norwegian Continental Margin: Sediments, Transport and Timing'. *Marine Geology*, 78, 77–107.

Jansen, J.H.F. 1976 'Late Pleistocene and Holocene history of the northern North Sea, based on acoustic reflection records'. *The Netherlands Journal of Sea Research*, 10, 1–43.

Jansen, J.H.F., Doppert, J.W.C., Hoogndoorn-Toering, K., de Jong, J. and G. Spaink 1979 'Late Pleistocene and Holocene deposits in the Witch and Fladen Ground area, northern North Sea'. *The Netherlands Journal of Sea Research*, 13, 1–39.

Jardine, W.G., Dickson, J.H., Haughton, P.D.W., Harkness, D.D., Bowen, D.Q. and G.A. Sykes 1988 'A Late Middle Devensian interstadial site at Sourlie, near Irvine, Strathclyde'. *Scottish Journal of Geology*, 24, 288–295.

Jarvis, G.T. and Peltier, W.R. 1982 'Mantle convection as a boundary layer phenomenon'. *Geophysical Journal of the Royal Astronomical Society*, 68, 389–424.

Johansen, J. 1985 'Pollen diagrams from the Shetland and the Faeroe Islands'. *New Phytologist*, 75, 369–387.

Johnsen, S.J., Dansgaard, W., Clausen, H.B. and C.C. Langway Jr 1972 'Oxygen isotope profiles through the Antarctic and Greenland Ice Sheets'. *Nature*, 235, 429–434.

Johnson, R.G. 1982 'Matuyama–Brunhes polarity reversal dated at 790,000 yr BP by marine–astronomical correlations'. *Quaternary Research*, 17, 135–147.

Johnson, W.H. 1986 'Stratigraphy and correlation of the glacial deposits of the Lake Michigan lobe prior to 14 ka BP'. In V. Sibrava, D.Q. Bowen and G.M. Richmond (eds) *Quaternary Glaciations in the Northern Hemisphere*, 17–22, Pergamon Press, Oxford, 514pp.

Jouzel, J. *et al.* 1989 'A companion of deep Antarctic ice cores and their implications for climate between 65,000 and 15,000 years ago'. *Quaternary Research*, 31, 135–150.

Karrow, P.F. 1984 'Quaternary stratigraphy and history, Great Lakes – St. Lawrence region'. In R.J. Fulton (ed.) *Quaternary stratigraphy of Canada – A Canadian contribution to IGCP Project 24*, Geological Survey of Canada, Paper 84-10, 137–153.

Karte, J. 1987 'Pleistocene periglacial conditions and geomorphology in north central Europe'. In J. Boardman (ed.) *Periglacial processes and landforms in Britain and Ireland*, 67–76, Cambridge University Press.

Kellogg, T.B. 1976 'Late Quaternary climatic changes: Evidence from cores of Norwegian and Greenland Seas'. In R.M. Cline and J.D. Hays (eds) *Investigation of Late Quaternary Paleo-oceanography and Paleoclimatology*, Geological Society of America Memoir 145, 77–110.

Kellogg, T.B. 1986 'Late Quaternary paleo-climatology and paleo-oceanography of the Labrador Sea and Baffin Bay: an alternative viewpoint'. *Boreas*, 15, 331–344.

Kellogg, T.B., Duplessy, J.C. and N.J. Shackleton 1978 'Planktonic foraminiferal and oxygen isotopic stratigraphy and paleoclimatology of Norwegian Sea deep-sea cores'. *Boreas*, 7, 61–73.

Kennett, J.P. and Huddlestun, P. 1972 'Late Pleistocene palaeoclimatology foraminiferal biostratigraphy and tephra chronology, western Gulf of Mexico'. *Quaternary Research*, 2, 38–69.

Kind, N.V. 1972 'Late Quaternary climatic changes and glacial events in the Old and New World – Radiocarbon chronology'. *International Geological Congress, 24th session, Section 12, Quaternary Geology, Montreal, 1972*, 55–61.

Kind, N.V. 1975 'Glaciations in the Verkhoyansk Mountains and their place in the radiocarbon geochronology of the Siberian Late Anthropogene'. *Biuletyn Peryglacjalny*, 24, 41–54.

Klassen, R.W. 1983 'Lake Agassiz and the Late Glacial History of Northern Manitoba'. In J.T. Teller and L. Clayton (eds) *Glacial Lake Agassiz*, 97–116, Geological Association of Canada Special Paper 26.

Koerner, R.M. 1977 'Devon Island Ice Cap: core stratigraphy and paleoclimate'. *Science*, 196, 15–18.

Kolstrup, E. 1980 'Climate and stratigraphy in Northwestern Europe between 30,000 BP and 13,000 BP with special reference to the Netherlands'. *Mededelingen Ryks Geologische Dienst*, 32–15, 181–253.

Krinsley, D.B. 1963 'Glacial Geology'. In W.E. Davies, Krinsley, D.B. and A.H. Nichols (eds) *Geology of the North Star Bugt area, Northwest Greenland, Meddelelser om Gronland*, 162, 48–66.

Kuhle, M. 1987 'The Pleistocene Glaciation of Tibet and its impact on the Global Climate'. *International Union for Quaternary Research Congress, Ottawa, Canada, 1987*, Programme Abstracts p. 204.

Kuhle, M. 1988 'Geomorphological findings on the build-up of Pleistocene Glaciation in southern Tibet and on the problem of Inland Ice – results of the Shisha Pangma and Mt Everest Expedition 1984'. *Geo Journal*, 17, (4), 457–511.

Kukla, G.J. 1987 'Loess stratigraphy in Central China'. *Quaternary Science Reviews*, 6, 191–219.

Kukla, G., Heller, F., Ming, L.X., Chun, X.T., Tung Sheng, L. and A.Z. Sheng 1988 'Pleistocene climates in China dated by magnetic susceptibility'. *Geology*, 16, 811–814.

Kutzbach, J.E. and Guetter, P.J. 1986 'The influence of changing orbital parameters and surface boundary conditions on climate simulations for the past 18,000 years'. *Journal of Atmospheric Sciences*, 43, 1726–1759.

Kutzbach, J.E. and Wright, H.E. 1985 'Simulation of the Climate of 18,000 years BP; results for the North American/North Atlantic/European sector and comparison with the geological record of North America'. *Quaternary Science Reviews*, 4, 147–187.

Kvasov, D.D. 1978 'The Barents Ice Sheet as a relay regulator of Glacial–Interglacial Alternation'. *Quaternary Research*, 9, 288–299.

Labeyrie, L.D., Pichon, J.J., Labracherie, M., Ippolito, P., Duprat, J. and J.C. Duplessy 1986 'Melting history of Antarctica during the past 60,000 years'. *Nature*, 322, 701–706.

Lamb, H.H. 1982 *Climate, History and the Modern World*. Methuen, London, 387pp.

Lambeck, K. 1990 'Glacial rebound, sea-level change and mantle viscosity'. *Quarterly Journal of the Royal Astronomical Society*, 31, 1–30.

Lambeck, K., Johnston, P. and M. Nakada 1990 'Holocene glacial rebound and sea-level change in northwestern Europe'. *Geophysical Journal International*, in press.

Larsen, E. and Sejrup, H.-P. 1990 'Weichselian land–sea interactions: Western Norway–Norwegian Sea'. *Quaternary Science Reviews*, 9, 85–98.

Larsen, E., Gulliksen, S., Lauritzen, S.-E., Lie, R., Lovlie, R. and J. Mangerud 1987 'Cave stratigraphy in western Norway: multiple Weichselian glaciations and interstadial vertebrate fauna'. *Boreas*, 16, 267–292.

LaSalle, P. and Elson, J.A. 1975 'Emplacement of the St. Narcisse moraine as a climatic event in eastern Canada'. *Quaternary Research*, 5, 621–625.

Leonard, A.B. and Frye, J.C. 1954 'Ecological conditions accompanying loess deposition in the Great Plains region'. *Journal of Geology*, 62, 399–404.

Livingstone, D.A. 1962 'Age of deglaciation in the Ruwenzori range, Uganda'. *Nature*, 194, 859–860.

Livingstone, D.A. 1980 'Environmental changes in the Nile headwaters'. In M.A.J. Williams and H. Faure (eds) *The Sahara and the Nile*, 339–360, Balkema, Rotterdam.

Long, D. and Morton, A.C. 1987 'An ash fall within the Loch Lomond Stadial'. *Journal of Quaternary Science*, 2, 97–102.

Long, D., Laban, C., Streif, H., Cameron, T.D.J. and R.T.E. Schuttenhelm 1988 'The sedimentary record and climatic variation in the southern North Sea'. *Philosophical Transactions of the Royal Society of London, Series B*, 318, 523–537.

Lorius, C.L., Merlivat, J., Jouzel, J. and M. Pourchet 1979 'A 30,000 year isotope climatic record from Antarctic ice'. *Nature*, 280, 644–648.

Lorius, C., Barkov, N.I., Jouzel, J., Korotkevich, Y.S., Kotlyakov, V.M. and D. Raynaud 1988 'Antarctic ice core: CO_2 and change over the last climatic cycle'. *EOS*, 69, 681–684.

Lowe, J.J. and Gray, J.M. 1980 'The stratigraphic subdivision of the Lateglacial of North-West Europe'. In J.J. Lowe, J.M. Gray and J.E. Robinson (eds) *Studies in the Lateglacial of North-West Europe*, 157–175, Pergamon, Oxford and New York.

Lowe, J.J. and Walker, M.J.C. 1984 *Reconstructing Quaternary Environments*. Longman Group Ltd, London, 389pp.

Lozano, J.A. and Hays, J.D. 1976 'Relationship of radiolarian assemblages to sediment types and physical oceanography in the Atlantic and western Indian Ocean sectors of the Antarctic Ocean'. In R.M. Cline and J.D. Hays (eds) *Investigation of Late Quaternary Paleoceanography and Paleoclimatology*, Geological Society of America Memoir 145, 303–336.

Lundqvist, J. 1986a 'Late Weichselian glaciation and deglaciation in Scandinavia'. In V. Sibrava, D.Q. Bowen and G.M. Richmond (eds) *Quaternary Glaciations in the Northern Hemisphere*, 269–292, Pergamon Press, Oxford, 514pp.

Lundqvist, J. 1986b 'Stratigraphy of the central area of the Scandinavian glaciation'. In V. Sibrava, D.Q. Bowen and G.M. Richmond (eds) *Quaternary Glaciations in the Northern Hemisphere*, 251–268, Pergamon Press, Oxford, 514pp.

Lundqvist, J. and Lagerback, R. 1976 'The Parve Fault; a late-glacial fault in the Precambrian of Swedish Lapland'. *Geologiska Foreningens Stockholm Forhandlinger*, 98, 45–51.

Maarleveld, G.C. 1976 'Periglacial phenomena and the mean annual temperature during the last glacial time in the Netherlands'. *Biuletyn Peryglacjalny*, 26, 57–78.

Manabe, S. and Hahn, D.G. 1977 'Simulation of the tropical climate of an Ice Age'. *Journal of Geophysical Research*, 82, 3889–3911.

Mangerud, J., Larsen, E., Longva, O. and E. Sonstegaard 1979 'Glacial history of western Norway 15,000–10,000 BP'. *Boreas*, 8, 179–187.

Mangerud, J., Lie, S.E., Furnes, H., Kristiansen, I.L. and L. Lomo 1984 'A Younger Dryas ash bed in Western Norway, and its possible correlations with tephra in cores from the Norwegian Sea and the North Atlantic'. *Quaternary Research*, 21, 85–104.

Marthinussen, M. 1960 'Coast and fjord area of Finmark. With remarks on some other districts'. *Norges Geologiske Undersokelse*, 208, 416–434.

Marthinussen, M. 1962 '[14]C-datings referring to shore lines, transgressions, and glacial substages in northern Norway'. *Norges Geologiske Undersokelse*, 213, 118–169.

Marthinussen, M. 1974 'Contributions to the Quaternary geology of north-easternmost Norway, and the closely adjoining foreign territories'. *Norges Geologiske Undersokelse*, 315, 157pp.

Martinson, D.G., Pisias, N.G., Hays, J.D., Imbrie, J., Moore, T.C. and N.J. Shackleton 1987 'Age, dating and orbital theory of the Ice Ages: Development of a high resolution 0 to 300,000-year chronostratigraphy'. *Quaternary Research*, 27, 1–29.

Mason, B.J. 1976 'Towards the understanding and prediction of climatic variation'. *Quarterly Journal of the Royal Meteorological Society*, 102, 473–498.

Mathews, J.V., Schweger, C.E. and O.L. Hughes 1989 'Climatic change in Eastern Beringia during oxygen isotope stages 2 and 3: proposed thermal events'. In L.D. Carter, T.D. Hamilton and J.D. Galloway (eds) *Late Cenozoic History of the Interior Basins of Alaska and the Yukon*, 34–38, US Geological Survey Circular 1026.

Matsch, C.L. 1983 'River Warren, the southern outlet to Glacial Lake Agassiz'. In J.T. Teller and L. Clayton (eds) *Glacial Lake Agassiz*, Geological Association of Canada Special Paper 26, 231–244.

Mayewski, P.A., Denton, G.H. and T.J. Hughes 1981 'Late Wisconsin Ice Sheets of North America'. In G.H. Denton and T.J. Hughes (eds) *The Last Great Ice Sheets*, 67–178, John Wiley and Sons, New York.

Mehringer, P.J., Shepard, J.C. and F.F. Foit 1984 'The age of Glacier Peak tephra in west-central Montana'. *Quaternary Research*, 21, 36–41.

Melchior, P. 1978 *The Tides of the Planet Earth*, Pergamon Press, Oxford.

Mercer, J.H. 1968 'Antarctic ice and Sangamon sea level'. *International Association of Scientific Hydrology Publication*, 79, 217–225.

Mercer, J.H. 1969 'The Alleröd oscillation: A European climatic anomaly?'. *Arctic and Alpine Research*, 1, 227–234.

Mercer, J.H. 1978 'West Antarctic Ice Sheet and CO_2 greenhouse effect: A threat of disaster'. *Nature*, 271, 321–325.

Mercer, J.H. 1983 'Cenozoic glaciation in the Southern Hemisphere'. *Annual Review of Earth and Planetary Science*, 11, 99–132.

Mercer, J.H. and Palacios, O. 1977 'Radiocarbon dating of the last glaciation in Peru'. *Geology*, 5, 600–604.

Mesolella, K.J., Mathews, R.K., Broecker, W.S. and D.L. Thurber 1969 'The astronomical theory of climatic change: Barbados data'. *Journal of Geology*, 77, 250–274.

Mickelson, D.M., Clayton, L., Fullerton, D.S. and H.W. Borns Jr 1983 'The Late Wisconsin Glacial Record of the Laurentide Ice Sheet in the United States'. In S.C. Porter (ed.) *Late-Quaternary Environments of the United States*, 3–37, Longman Group Ltd, London.

Milankovitch, M.M. 1941 *Canon of insolation and the ice-age problem*, Koniglich

Serbische Akademie, Belgrade. English translation by the Israel Program for Scientific Translations, published for the US Department of Commerce, and the National Science Foundation, Washington, DC (1969).

Mitchell, G.F., Penny, L.F., Shotton, F.W. and R.G. West 1973 'A correlation of Quaternary deposits in the British Isles'. *Geological Society of London Special Report*, 4, 1–99.

Mook, W. and Woillard, G. 1982 'Carbon-14 dates at Grand Pile. Correlation of land and sea chronologies'. *Science*, 215, 159–161.

Mörner, N.-A. 1971 'Eustatic changes during the last 20,000 years and a method of separating the isostatic and eustatic factors in an uplifted area'. *Palaeogeography, Palaeoclimatology and Palaeoecology*, 9, 153–181.

Mörner, N.-A. 1976 'Eustasy and geoid changes'. *Journal of Geology*, 84, 123–152.

Mörner, N.-A. 1980a 'The Fennoscandian uplift: geological data and their geodynamical implication'. In N.-A. Mörner (ed.) *Earth Rheology, Isostasy and Eustasy*, 251–284, John Wiley and Sons, Chichester, 599pp.

Mörner, N.-A. 1980b 'Eustasy and geoid changes as a function of core/mantle changes'. In N.-A. Mörner (ed.) *Earth Rheology, Isostasy and Eustasy*, 535–554, John Wiley and Sons, Chichester, 599pp.

Mörner, N.-A. 1987 'Models of global sea level changes'. In M.J. Tooley and I. Shennan (eds) *Sea Level Changes*, 332–355, Basil Blackwell, Oxford, 397pp.

Mörner, N.-A. and Dreimanis, A. 1973 'The Erie interstade'. In R.F. Black, R.P. Goldthwait and H.B. Willman (eds) *The Wisconsinan Stage*, Geological Society of America Memoir 136, 107–134.

Mott, R.J., Grant, D.R., Stea, R. and S. Ochietti 1986 'Late-glacial climatic oscillation in Atlantic Canada equivalent to the Alleröd/Younger Dryas event'. *Nature*, 323, 247–250.

Muller, S.W. 1947 'Permafrost or permanently frozen ground and related engineering problems'. J.W. Edwards, Ann Arbor, Michigan, 231pp.

Mullineaux, D.R., Waldron, H.H. and M. Rubin 1965 'Stratigraphy and chronology of late interglacial and early Vashon glacial time in the Seattle area, Washington'. *United States Geological Survey Bulletin*, 1194-D, 10pp.

Nicholson, S. and Flohn, H. 1980 'African environmental and climatic changes and the general circulation in late Pleistocene and Holocene'. *Climatic Change*, 2, 313–48.

Ninkovich, D., Shackleton, N.J., Abdel-Monem, A.A., Obradovich, J.D. and G. Izett 1978 'K–Ar age of the late Pleistocene eruption of Toba, north Sumatra'. *Nature*, 276, 574–577.

Olausson, E. 1965 'Evidence of climatic changes in deep-sea cores, with remarks on isotopic temperature analysis'. *Progress in Oceanography*, 3, 221–252.

Omoto, K. 1977 'Geomorphic development of the Soya Coast, East Antarctica'. *Science Reports of the Tohoku University, 7th Series (Geography)*, 27, 95–148.

Ota, Y. 1986 'Marine terraces as reference surfaces in late Quaternary tectonic studies: examples from the Pacific Rim'. *Royal Society of New Zealand*, 24, 357–374.

Ota, Y. 1987 'Sea-level changes during the Holocene: the Northwest Pacific'. In R.J.N. Devoy (ed.) *Sea Surface Studies – a Global View*, 348–374, Croom Helm Ltd. London.

Ota, Y. and Machida, H. 1987 'Quaternary sea level changes in Japan'. In M.J. Tooley and I. Shennan (eds) *Sea Level Changes*, 182–224, Basil Blackwell, Oxford.

Parkin, D.W. and Shackleton, N.J. 1973 'Trade winds and temperature correlations down a deep-sea core off the Saharan coast'. *Nature*, 245, 455–457.

Paskoff, R.P. 1977 'Quaternary of Chile: The state of research'. *Quaternary Research*, 8, 2–31.

Patton, P.C. and Schumm, S.A. 1981 'Ephemeral stream processes: implications for studies of Quaternary valley fills'. *Quaternary Research*, 15, 24–43.

Peltier, W.R. 1981 'Ice age geodynamics', *Annual Review Earth Planetary Science*, 9, 199–225.

Peltier, W.R. and Andrews, J.T. 1983 'Glacial geology and glacial isostasy of the Hudson

Bay region'. In D.E. Smith and A.G. Dawson (eds) *Shorelines and Isostasy*, 285–320, Academic Press, London.

Penck, A. and Bruckner, E. 1909 *Die Alpen im Eiszeitalter*, Tauchnitz, Leipzig, 1199pp.

Persson, C. 1971 'Tephrochronological investigation of peat deposits in Scandinavia and on the Faeroe Islands'. *Sveriges Geologiska Undersokning*, Arbok 65, 2, 1–34.

Petit, J.R., Mounier, L., Jouzel, J., Korotkevich, Y.S., Kotlyakov, V.I. and C. Lorius 1990 'Palaeoclimatological and chronological implications of the Vostok core dust record'. *Nature*, 343, 56–58.

Pewe, T.L. 1975 *Quaternary Geology of Alaska*, Geological Survey Professional Paper 835, 145pp.

Pewe, T.L. 1983 'The periglacial environment in North America during Wisconsin time'. In S.C. Porter (ed.) *Late Quaternary Environments of the United States, Volume I, The Late Pleistocene*, 157–189, Longman Group Ltd, London.

Piper, D.J.W., Mudie, P.J., Letson, J.R., Barnes, N.E. and R.J. Iuliucci 1986 'The marine geology of the inner Scotian Shelf off the south shore, Nova Scotia'. *Geological Survey of Canada. Paper* 85-19, 65pp.

Pirazzoli, P. 1987 'Sea-Level changes in the Mediterranean'. In M.J. Tooley and I. Shennan (eds) *Sea Level Changes*, 152–181, Basil Blackwell, Oxford.

Pissart, A. 1987 'Weichselian periglacial structures and their environmental significance: Belgium, the Netherlands and northern France'. In J. Boardman (ed.) *Periglacial processes and landforms in Britain and Ireland*, 77–88, Cambridge University Press.

Porter, S.C. 1975 'Equilibrium-line altitudes of Late Quaternary glaciers in the Southern Alps, New Zealand'. *Quaternary Research*, 5, 27–47.

Porter, S.C. 1978 'Glacier Peak tephra in the North Cascade Range, Washington: stratigraphy, distribution, and relationship to Late-Glacial Events'. *Quaternary Research*, 10, 30–41.

Porter, S.C. 1979 'Hawaiian glacial ages'. *Quaternary Research*, 12, 161–87.

Porter, S.C. 1981 'Pleistocene glaciation in the southern Lake District of Chile'. *Quaternary Research*, 16, 263–292.

Porter, S.C., Pierce, K.L. and T.D. Hamilton 1986 'Late Wisconsin mountain glaciation in the Western United States'. In S.C. Porter (ed.) *Late-Quaternary Environments of the United States*, 71–114, Longman Group Ltd, London.

Prest, V.K. and Grant, D.R. 1969 'Retreat of the last ice sheet from the Maritime Provinces-Gulf of St. Lawrence region'. *Geological Survey of Canada, Paper* 69-23, 15pp.

Price, R.J. 1983 *Scotland's Environment during the last 30,000 years*. Scottish Academic Press, Edinburgh, 224pp.

Pye, K. 1984 'Loess'. *Progress in Physical Geography*, 8, 176–217.

Pye, K. 1987 *Aeolian dust and dust deposits*, Academic Press, London.

Rabek, K., Ledbetter, M.T. and Williams, D.F. 1985 'Tephrochronology of the Western Gulf of Mexico for the last 185,000 years'. *Quaternary Research*, 23, 403–416.

Raisbeck, G.M., Yiou, F., Bourles, D., Lorius, C., Jouzel, J. and N.I. Barkov. 1987 'Evidence for two intervals of enhanced [10]Be deposition in Antarctic ice during the last glacial period'. *Nature*, 326, 273.

Rampino, M.R., Self, S. and R.W. Fairbridge 1979 'Can rapid climatic change cause volcanic eruptions?'. *Science*, 206, 826– 829.

Reger, R.D. and Pewe, T.L. 1976 'Cryoplanation terraces: Indicators of a permafrost environment'. *Quaternary Research*, 6, 99–109.

Richard, S.H. 1978 'Age of the Champlain Sea and "Lampsilis Lake" Episode in the Ottawa–St. Lawrence Lowlands'. *Geological Survey of Canada, Paper* 78-1C, 23–28.

Richmond, G.M. 1986a 'Stratigraphy and correlation of glacial deposits of the Rocky Mountains, the Colorado Plateau and the Ranges of the Great Basin'. In V. Sibrava, D.Q. Bowen and G.M. Richmond (eds) *Quaternary Glaciations in the Northern*

Hemisphere, 99–128, Pergamon Press, Oxford, 514pp.

Richmond, G.M. 1986b 'Stratigraphy and Chronology of Glaciations in Yellowstone National Park'. In V. Sibrava, D.Q. Bowen and G.M. Richmond (eds) *Quaternary Glaciations in the Northern Hemisphere*, 83–98, Pergamon Press, Oxford, 514pp.

Richmond, G.M. and Fullerton, D.S. 1986a 'Introduction to Quaternary Glaciations in the United States of America'. In V. Sibrava, D.Q. Bowen and G.M. Richmond (eds) *Quaternary Glaciations in the Northern Hemisphere*, 3–10, Pergamon Press, Oxford, 514pp.

Richmond, G.M. and Fullerton, D.S. 1986b 'Summation of Quaternary Glaciations in the United States of America'. In V. Sibrava, D.Q. Bowen and G.M. Richmond (eds) *Quaternary Glaciations in the Northern Hemisphere*, 183–196, Pergamon Press, Oxford, 514pp.

Rind, D., Peteet, D., Broecker, W.G., McIntyre, A. and W. Ruddiman 1986 'The impact of cold North Atlantic sea surface temperatures on climate: implications for the Younger Dryas cooling (11-10k)'. *Climate Dynamics*, 1, 3–33.

Ringrose, P.S. 1989 'Palaeoseismic(?) liquefaction event in late Quaternary lake sediment at Glen Roy, Scotland'. *Terra Nova*, 1, 1, 57–62.

Roberts, N. 1989 *The Holocene*, Basil Blackwell, Oxford, 227pp.

Rognon, P. (ed.) 1976 'Oscillations climatiques au Sahara depuis 40,000 ans'. *Revue de Geographie Physique et de Geologie Dynamique*, 18, 147–282.

Rognon, P. 1980 'Fluviatile piedmont deposits'. In M.A.J. Williams and H. Faure (eds) *The Sahara and the Nile*, 118–132, Balkema, Rotterdam.

Rose, W.I. and Chesner, C.A. 1987 'Dispersal of ash in the great Toba eruption, 75ka'. *Geology*, 15, 913–917.

Rose, W.I. Jr, Grant, N.K. and Easter, J. 1979 'Geochemistry of the Los Chocoyos Ash, Quezaltenango Valley, Guatemala'. In W. Elston and C. Chapin (eds) *Ash Flow Tuffs*, Geological Society of America Special Paper 180, 87–100.

Rossignol-Strick, M. and Duzer, D. 1980 'Late Quaternary West African climate inferred from palynology of Atlantic deep-sea cores'. *Palaeoecology of Africa*, 12, 227–228.

Ruddiman, W.F. 1977 'Late Quaternary deposition of ice-rafted sand in the subpolar North Atlantic (lat 40 to 65°N)'. *Geological Society of America Bulletin*, 88, 1813–1827.

Ruddiman, W.F. and Glover, L.K. 1972 'Vertical mixing of ice-rafted volcanic ash in North Atlantic sediments'. *Geological Society of America Bulletin*, 83, 2817–2836.

Ruddiman, W.F. and Glover, L.K. 1975 'Subpolar North Atlantic circulation at 9,300 yr BP: faunal evidence'. *Quaternary Research*, 5, 361–389.

Ruddiman, W.F. and McIntyre, A. 1973 'Time-transgressive deglacial retreat of polar waters from the North Atlantic'. *Quaternary Research*, 3, 117–130.

Ruddiman, W.F. and McIntyre, A. 1981a 'The mode and mechanism of the last deglaciation: oceanic evidence'. *Quaternary Research*, 16, 125–134.

Ruddiman, W.F. and McIntyre, A. 1981b 'The North Atlantic Ocean during the last deglaciation'. *Palaeogeography, Palaeoclimatology and Palaeoecology*, 35, 145–214.

Ruddiman, W.F. and McIntyre, A. 1981c 'Oceanic mechanisms for amplification of the 23,000 year ice-volume cycle'. *Science*, 212, 617–627.

Ruddiman, W.F., McIntyre, A., Niebler-Hunt, V. and J.T. Durazzi 1980 'Oceanic evidence for the mechanism of rapid northern hemisphere glaciation'. *Quaternary Research*, 13, 33–64.

Ruhe, R.V. 1986 'Depositional environment of Late Wisconsin loess in the Midcontinental United States'. In S.C. Porter (ed.) *Late-Quaternary environments of the United States*, 130–137, Longman Group Ltd, London, 406pp.

Russell, I.C. 1885 *Geological History of Lake Lahontan, a Quaternary Lake of Northwestern Nevada*, United States Geological Survey Monograph 11.

Russell, R.J. 1944 'Lower Mississippi Valley loess'. *Geological Society of America Bulletin*, 55, 1–40.

Rutter, N.W. 1984 'Pleistocene history of the western Canadian ice-free corridor'. In R.J. Fulton (ed.) *Quaternary Stratigraphy of Canada – A Canada Contribution to IGCP Project 24*, Geological Survey of Canada, Paper 84-10.

Sakaguchi, Y. 1978 'Climatic changes in central Japan since 38,400 yr BP'. *Bulletin of the Department of Geography, University of Tokyo*, 10, 1–10.

Sancetta, C., Imbrie, J. and N.G. Kipp 1973 'Climatic Record of the Past 130,000 Years in North Atlantic Deep-Sea Core V23-82: correlation with the terrestrial record'. *Quaternary Research*, 3, 110–116.

Sarnthein, M. 1978 'Sand deserts during Glacial Maximum and climatic optimum'. *Nature*, 273, 43–46.

Sarnthein, M. and Koopman, B. 1980 'Late Quaternary deep-sea record of north-west African dust supply and wind circulation'. *Palaeoecology of Africa*, 12, 238–253.

Schumm, S.A. 1965 'Quaternary palaeohydrology'. In H.E. Wright and D.G. Frey (eds) *The Quaternary of the United States*, 783-794, Princeton University Press, Princeton, New Jersey.

Schumm, S.A. 1974 'Geomorphic thresholds and complex response of drainage systems'. In M.E. Morisawa (ed.) *Fluvial Geomorphology*, Publications in Geomorphology, SUNY-Binghamton, 299–310.

Schumm, S.A. 1977 *The Fluvial System*, John Wiley and Sons, Chichester, 356pp.

Schweger, C.E. and Matthews, J.V. Jr 1985 'Early and middle Wisconsinan environments of eastern Beringia: Stratigraphic and palaeoecological implications of the Old Crow Tephra'. *Geographie Physique et Quaternaire*, 39, 275–290.

Sejrup, H.P. 1987 'Molluscan and foraminiferal biostratigraphy of an Eemian–Early Weichselian section on Karmøy, southwestern Norway'. *Boreas*, 16, 27–42.

Sejrup, H.P., Aarseth, I., Ellingsen, K.L., Reither, E., Jansen, E., Lovlie, R., Bent, A., Brigham-Grette, J., Larsen, E. and M. Stoker 1987 'Quaternary stratigraphy of the Fladen area, central North Sea: a multidisciplinary study', *Journal of Quaternary Science*, 2, 35–58.

Sejrup, H.-P., Sjoholm, J., Furnes, H., Beyer, I., Eide, L., Jansen, E. and J. Mangerud 1989 'Quaternary tephrachronology on the Iceland Plateau, north of Iceland'. *Journal of Quaternary Science*, 4, 109–114.

Servant, M. and Servant-Valdary, S. 1980 'L'environement Quaternaire du bassin du Tchad'. In M.A.J. Williams and H. Faure (eds) *The Sahara and the Nile*, 133–162, Balkema, Rotterdam.

Settle, M. 1978 'Volcanic eruption clouds and the thermal power output of explosive eruptions'. *Journal of Volcanology Geothermal Research*, 3, 1727–1739.

Shackleton, N.J. 1967 'Oxygen isotope analyses and Pleistocene temperatures reassessed'. *Nature*, 215, 15–17.

Shackleton, N.J. 1987 'Oxygen isotopes, ice volume and sea level'. *Quaternary Science Reviews*, 6, 183–190.

Shackleton, N.J. and Opdyke, N.D. 1973 'Oxygen isotope and palaeomagnetic stratigraphy of equatorial Pacific core V28-238: oxygen isotope temperatures and ice volumes on a 10^5 and 10^6 year scale'. *Quaternary Research*, 3, 39–55.

Shi, Y., Ren, B., Wang, J. and E. Derbyshire 1986 'Quaternary Glaciation in China'. In V. Sibrava, D.Q. Bowen and G.M. Richmond (eds) *Quaternary Glaciations in the Northern Hemisphere*, 503–507, Pergamon Press, Oxford, 514pp.

Shilts, W.W. 1980 'Flow patterns in the central North American ice sheet'. *Nature*, 286, 213–218.

Shilts, W.W. 1984 'Quaternary events, Hudson Bay Lowland and southern district of Keewatin'. In R.J. Fulton (ed.) *Quaternary stratigraphy of Canada – A Canadian Contribution to IGCP Project 24*, Geological Survey of Canada, Paper 84-10, 117–126.

Shilts, W.W., Cunningham, C.M. and C.A. Kaszycki 1979 'Keewatin Ice Sheet – Re-evaluation of the traditional concept of the Laurentide Ice Sheet.' *Geology*, 7, 537–541.

Shotton, F.W. 1986 'Glaciations in the United Kingdom'. In V. Sibrava, D.Q. Bowen and G.M. Richmond (eds) *Quaternary Glaciations in the Northern Hemisphere*, 293–298, Pergamon Press, Oxford, 514pp.

Singh, G. and Geissler, E.A. 1985 'Late Cainozoic history of vegetation, fire, lake levels and climate, at Lake George, New South Wales, Australia'. *Philosophical Transactions of the Royal Society of London*, B311, 379–447.

Sissons, J.B. 1967 *The Evolution of Scotland's Scenery*, Oliver and Boyd, Edinburgh.

Sissons, J.B. 1972 'Dislocation and non-uniform uplift of raised shorelines in the western part of the Forth valley'. *Transactions Institute of British Geographers*, 55, 145–149.

Sissons, J.B. 1974a 'Late-glacial marine erosion in Scotland'. *Boreas*, 3, 41–48.

Sissons, J.B. 1974b 'The Quaternary in Scotland: a review'. *Scottish Journal of Geology*, 10, 311–337.

Sissons, J.B. 1979a 'The Loch Lomond Stadial in the British Isles'. *Nature*, 280, 199–202.

Sissons, J.B. 1979b 'Catastrophic lake drainage in Glen Spean and the Great Glen, Scotland'. *Journal of the Geological Society of London*, 136, 215–224.

Sissons, J.B. 1981a 'The Last Scottish ice sheet: facts and speculative discusssion'. *Boreas*, 10, 1–17.

Sissons, J.B. 1981b 'Lateglacial marine erosion and a jökulhlaup deposit in the Beauly Firth'. *Scottish Journal of Geology*, 17, 7–19.

Sissons, J.B. 1983 'Shorelines and Isostasy in Scotland'. In D.E. Smith and A.G. Dawson (eds) *Shorelines and Isostasy*, 209–226, Academic Press, London.

Sissons, J.B. and Cornish, R. 1982 'Differential glacio-isostatic uplift of crustal blocks at Glen Roy, Scotland'. *Quaternary Research*, 18, 268–288.

Sissons, J.B. and Walker, M.J.C. 1974 'Lateglacial site in the central Grampian Highlands'. *Nature*, 249, 822–824.

Sissons, J.B., Cullingford, R.A. and D.E. Smith 1966 'Late-glacial and post-glacial shorelines in South-East Scotland'. *Transactions of the Institute of British Geographers*, 39, 9–18.

Skinner, B.J. and Porter, S.C. 1987 *Physical Geology*, John Wiley and Sons, New York, 750pp.

Smalley, I.J. and Smalley, V. 1983 'Loess material and loess deposits: formation, distribution and consequences'. In M.E. Brookfield and T.S. Ahlbrandt (eds) *Eolian and Sediments and Processes*, 51–68, Elsevier, Amsterdam.

Smith, G.I. and Street-Perrott, F.A. 1983 'Pluvial lakes of the Western United States'. In S.C. Porter (ed.) *Late-Quaternary Environments of the United States, Volume 1, The Late Pleistocene*, 190–214, Longman Group Ltd, London.

Sollid, J.L., Andersen, S., Hamre, N., Kjeldsen, O., Selvigsen, O., Sturod, S., Tveitera, T. and A. Wilhelmsen 1973 'Deglaciation of Finmark, North Norway'. *Norsk Geografisk Tidsskrift*, 27, 233–325.

St Onge, D.A. 1987 'The Sangamonian Stage and the Laurentide Ice Sheet'. *Geographie Physique et Quaternaire*, 41, 189–198.

Starkel, L. 1983 'The reflection of hydrologic changes in the fluvial environment of the temperate zone during the last 15,000 years'. In K.J. Gregory (ed.) *Background to Palaeohydrology: a Perspective*, 213–236, John Wiley and Sons, Chichester.

Starkel, L. 1987 'The evolution of European Rivers – a complex response'. In K.J. Gregory, J. Lewin and J.B. Thornes (eds) *Palaeohydrology in Practice*, 333–340, John Wiley and Sons, Chichester.

Starkel, L. 1990 'Global continental paleohydrology'. *Palaeogeography, Palaeoclimatology, Palaeoecology (Global and Planetary Change Section)*, 82, 73–77.

Stauffer, P.H., Nishimura, S. and B.C. Batchelor 1980 'Volcanic ash in Malaya from a catastrophic eruption of Toba, Sumatra, 30,000 years ago'. In S. Nishimura (ed.) *Physical Geology of Indonesian Island Arcs*, Kyoto, 156–164.

Stoker, M.S., Long, D. and J.A. Fyfe 1985 'The Quaternary succession in the central North Sea'. *Newsletters on Stratigraphy*, 14, 119–128.

Street, F.A. 1981 'Tropical palaeoenvironments'. *Progress in Physical Geography*, 5, 157–185.

Street, F.A. and Grove, A.T. 1979 'Global maps of lake-level fluctuations since 30,000 yr BP'. *Quaternary Research*, 12, 83–118.

Street-Perrott, F.A. and Perrott, R.A. 1990 'Abrupt climate fluctuations in the tropics: the influence of Atlantic Ocean circulation'. *Nature*, 343, 607–612.

Stuiver, M., Denton, G.H., Hughes, T.J. and J.L. Fastook 1981 'History of the Marine Ice Sheet in West Antarctica during the last glaciation: a working hypothesis'. In G.H. Denton and T.J. Hughes (eds) *The Last Great Ice Sheets*, 319–439, John Wiley and Sons, New York.

Suggate, R.P. and Moar, N.T. 1970 'Revision of the chronology of the Late Otiran glacial'. *New Zealand Journal of Geology and Geophysics*, 13, 742–746.

Sutherland, D.G. 1981 'The high-level marine shell beds of Scotland and the build-up of the last Scottish ice sheet'. *Boreas*, 10, 247–254.

Sutherland, D.G. 1984 'The Quaternary deposits and landforms of Scotland and the neighbouring shelves: a review'. *Quaternary Science Reviews*, 3, 157–254.

Talbot, M.R. 1980 'Environmental responses to climatic change in the West African Sahel over the past 20,000 years'. In M.A.J. Williams and H. Faure (eds) *The Sahara and the Nile*, 37–62, Balkema, Rotterdam.

Tanner, V. 1930 'Studier over kvartarsystemet i Fennoskandias Nordliga Delar'. *Fennia*, 53, 1, 484–537, Helsinki-Helsingfors.

Tedrow, J.F. 1970 'Soil investigations in Inglefield Land, Greenland'. *Meddelelser om Gronland*, 188, 93pp.

Teller, J.T. 1989 'Importance of the Rossendale site in establishing a deglacial chronology along the Southwestern margin of the Laurentide ice sheet'. *Quaternary Research*, 32, 12–23.

Teller, J.T. 1990 'Meltwater and precipitation runoff to the North Atlantic, Arctic, and Gulf of Mexico from the Laurentide ice sheet and adjacent regions during the Younger Dryas'. *Paleoceanography*, 5, 897–905.

Teller, J.T. and Clayton, L. (eds) 1983 *Glacial Lake Agassiz*, Geological Association of Canada Special Paper 26, 451pp.

Teller, J.T. and Mahnic, P. 1988 'History of sedimentation in the northwestern Lake Superior basin and its relation to Lake Agassiz overflow'. *Canadian Journal of Earth Science*, 25, 1660–1673.

Teller, J.T. and Thorleifson, L.H. 1983 'The Lake Agassiz–Lake Superior Connection'. In J.T. Teller and L. Clayton (eds) *Glacial Lake Agassiz*, 261–290, Geological Association of Canada Special Paper 26.

Ten Brink, N.V. and Weidick, A. 1973 'Greenland ice sheet history since the last glaciation'. *Quaternary Research*, 4, 429–440.

Thiede, J. 1974 'A Glacial Mediterranean'. *Nature*, 276, 680–683.

Thomas, D.S.G. 1989 'Reconstructing ancient arid environments'. In D.S.G. Thomas (ed.) *Arid Zone Geomorphology*, 311–334, Belhaven Press and Halsted Press, London and New York, 372pp.

Thomas, D.S.G. and Goudie, A.S. 1984 'Ancient ergs of the southern hemisphere'. In J.C. Vogel (ed.) *Late Cainozoic palaeoclimates of the southern hemisphere*, 407–418, Balkema, Rotterdam.

Thompson, R. and Bradshaw, R.H.W. 1986 'The distribution of ash in Icelandic lake sediments and the relative importance of mixing and erosion processes'. *Journal of Quaternary Science*, 1, 3–11.

Thornes, J.B. 1987 'Models for Palaeohydrology in Practice'. In K.J. Gregory, J. Lewin and J.B. Thornes (eds) *Palaeohydrology in Practice: a river basin analysis*, 17–36, John Wiley and Sons, Chichester.

Thorson, R.M. and Hamilton, T.D. 1986 'Glacial geology of the Aleutian Islands (based on contributions of Robert F. Black)'. In T.D. Hamilton, K.M. Reed and R.M. Thorson (eds) *Glaciation in Alaska*, Alaska Geological Society, 171–192.

Thunell, R., Federman, A., Sparks, S. and D. Williams 1979 'The age, origin and volcanological significance of the Y-5 ash layer in the Mediterranean'. *Quaternary Research*, 12, 241–253.

Tricart, J. 1974 'Existence des periodes seches au Quaternaire en Amazonie et dans les regions voisines'. *Revue de Geomorphologie Dynamique*, 23, 145–158.

Tricart, J. 1975 'Existence de periode seche au Quaternaire en Amazonie et dans les regions voisins'. *Revue de Geomorphologie Dynamique*, 4, 145–158.

Tungsheng, L., Shouxin, Z. and Jiaomao, H. 1986 'Stratigraphy and palaeoenvironmental changes in the loess of Central China'. In V. Sibrava, D.Q. Bowen and G.M. Richmond (eds) *Quaternary Glaciations in the Northern Hemisphere*, 489–495, Pergamon Press, Oxford, 514pp.

Van Bemmelen, R.W. 1949 *The Geology of Indonesia, Volume 1A: general geology of Indonesia and adjacent archipelagos*, Martinus Nijhoff, The Hague.

Van der Hammen, T. 1974 'The Pleistocene changes of vegetation and climate in tropical South America'. *Journal of Biogeography*, 1, 3–26.

Van der Hammen, T., Wijmstra, T.A. and W.H. Zagwijn 1971 'The floral record of the late Cenozoic in Europe'. In K.K. Turekian (ed.) *The Late Cenozoic Glacial Ages*, 391–424, Yale University Press, New Haven, Connecticut.

Van der Hammen, T., Barelds, J., De Jong, M. and A.A. De Veer 1981 'Glacial sequence and environmental history in the Sierra Nevada del Cocuy, Colombia'. *Palaeogeography, Palaeoclimatology, Palaeoecology*, 32, 247–340.

Van de Plassche, O. 1982 'Sea level change and water-level movements in the Netherlands during the Holocene'. *Mededelingen Rijks Geologische Dienst*, 36-1, 93pp.

Velichko, A.A. (ed.) 1984 *Late Quaternary Environments of the Soviet Union*, Longman Group Ltd, London, 327pp.

Velichko, A.A. 1984 'Introduction'. In A.A. Velichko (ed.) *Late Quaternary Environments of the Soviet Union*, xxiii–xxvii, Longman Group Ltd, London, 327pp.

Velichko, A.A. and Faustova, M.A. 1986 'Glaciations in the East European region of the USSR'. In V. Sibrava, D.Q. Bowen and G.M. Richmond (eds) *Quaternary Glaciations in the Northern Hemisphere*, 447–462, Pergamon Press, Oxford, 514pp.

Velichko, A.A. and Nechayev, V.P. 1984 'Late Pleistocene Permafrost in European USSR'. In A.A. Velichko (ed.) *Late Quaternary Environments of the Soviet Union*, 79–86, Longman Group Ltd, London, 327pp.

Velichko, A.A., Isayeva, L.L., Makeyev, V.M., Matishov, G.G. and M.A. Faustova 1984 'Late Pleistocene Glaciation of the Arctic Shelf, and the reconstruction of Eurasian ice sheets'. In A.A. Velichko (ed.) *Late Quaternary Environments of the Soviet Union*, 35–44, Longman Group Ltd, London.

Vernal, A. de and Hillaire-Marcel, C. 1987 'Paleo-environments along the eastern Laurentide Ice Sheet margin and timing of the last ice maximum and retreat'. *Geographie Physique et Quaternaire*, 41, 265–278.

Vernekar, A.D. 1972 'Long-period global variations of incoming solar radiation'. *Meteorological Monograph*, 12, (34).

Verstappen, H.Th. 1970 'Aeolian geomorphology of the Thar Desert and palaeoclimates'. *Zeitschrift für Geomorphologie*, Supplement Band 10, 104–120.

Vincent, J.-S. 1978 'Limits of ice advance, glacial lakes, and marine transgressions on Banks Island, District of Franklin: A preliminary interpretation'. *Geological Survey of Canada, Paper* 78-1C, 53–62.

Vincent, J.-S. 1984 'Quaternary stratigraphy of the western Canadian Arctic Archipelago'. In R.J. Fulton (ed.) *Quaternary stratigraphy of Canada – A Canadian Contribution to IGCP Project 24*, Geological Survey of Canada, Paper 84-10, 87–100.

Vincent, J.-S. and Hardy, L. 1979 'The evolution of Glacial Lakes Barlow and Ojibway, Quebec and Ontario'. *Geological Survey of Canada, Bulletin* 316, 18pp.

Vincent, J.-S. and Klassen, R.W. 1989 'Quaternary Geology of the Canadian Interior Plains'. In R.J. Fulton (ed.) *Quaternary Geology of Canada and Greenland. Geological Survey of Canada, Geology of Canada, no. 1* (Geological Society of America, The Geology of North America v. K-1), 98–174.

Vincent, J.-S. and Prest, V.K. 1987 'The Early Wisconsin history of the Laurentide Ice Sheet'. *Geographie Physique et Quaternaire*, 41, 199–214.

Vita-Finzi, C. 1986 *Recent Earth Movements: an introduction to neotectonics*, Academic Press, London, 226pp.

Vorren, K.D. 1978 'Late and middle Weichselian stratigraphy of Andoya, north Norway'. *Boreas*, 7, 19–38.

Vorren, T.O., Vorren, K.D., Torbjorn, A., Gulliksen, S. and R. Lovlie 1988 'The last deglaciation (20,000 to 11,000 BP) on Andoya, northern Norway'. *Boreas*, 17, 41–77.

Waitt, R.B. Jr 1980 'About forty last glacial Lake Missoula jökulhlaups through southern Washington'. *Journal of Geology*, 88, 653–679.

Waitt, R.B. Jr 1985 'Case for periodic, colosal jökulhlaups from Pleistocene glacial Lake Missoula'. *Geology Society of American Bulletin*, 1271–1286.

Walcott, R.I. 1970 'Isostatic response to loading of the crust in Canada'. *Canadian Journal of Earth Sciences*, 7, 716–726.

Walcott, R.I. 1972 'Past sea levels, eustasy, and deformation of the Earth'. *Quaternary Research*, 2, 1–14.

Walker, H.J. 1973 'Morphology of the North Slope'. In M.E. Britton (ed.) *Alaskan Arctic Tundra (Naval Arctic Research Laboratory, 25th Anniversary Celebration Proceedings)*, Arctic Institute of North America Technical Paper 25, 49–92.

Washburn, A.L. 1979 *Geocryology*, Edward Arnold, London.

West, R.G. 1972 *Pleistocene Geology and Biology*, Longman Group Ltd, London.

Westgate, J.A., Hamilton, T. and M.P. Gorton 1983 'Old Crow Tephra: a new Late Pleistocene stratigraphic marker across North-Central Alaska and Western Yukon Territory'. *Quaternary Research*, 19, 38–54.

Williams, M.A.J. and Adamson, D.A. 1980 'Late Quaternary depositional history of the Blue and White Nile rivers in central Sudan'. In M.A.J. Williams and H. Faure (eds) *The Sahara and the Nile*, 281–304, Balkema, Rotterdam.

Williams, R.B.G. 1975 'The British climate during the last glaciation: an interpretation based on periglacial phenomena'. In A.E. Wright and F. Moseley (eds) *Ice Ages: Ancient and Modern*, 95–120, Seel House Press, Liverpool.

Wilson, A.T. 1964 'Origin of ice ages: an ice shelf theory for Pleistocene glaciation'. *Nature*, 201, 147–149.

Wilson, L., Sparks, R.J.S., Huang, T.C. and N.D. Watkins 1978 'The control of volcanic column eruption heights by eruption energetics and dynamics'. *Journal of Geophysical Research*, 83, 1829–1836.

Woillard, G.M. 1978 'Grand Pile peat bog: a continuous pollen record for the past 140,000 years'. *Quaternary Research*, 9, 1–21.

Worzel, J.L. 1959 'Extensive deep-sea sub-bottom reflections identified as white ash'. *Proceedings of the National Academy of Science*, 45, 349–355.

Wright, H.E. Jr 1989 'The Amphi-Atlantic Distribution of the Younger Dryas Palaeoclimatic Oscillation'. *Quaternary Science Reviews*, 8, 295–306.

Wright, H.E. Jr and Barnowsky, C.W. 1984 'Introduction to the English Edition'. In A.A. Velichko (ed.) *Late Quaternary Environments of the Soviet Union*, xiii–xxii, Longman Group Ltd, London.

Wu, P. and Peltier, W.R. 1983 'Glacial isostatic adjustment and the free air gravity anomaly as a constraint upon deep mantle viscosity'. *Geophysical Journal of the Royal Astronomical Society*, 74, 377–449.

Wyrwoll, K. and Milton, D. 1976 'Widespread late Quaternary aridity in western Australia'. *Nature*, 264, 429–430.

Yonekura, N. and Ota, Y. 1986 'Sea-level changes and tectonics in the Late Quaternary'. *Recent Progress of Natural Sciences in Japan, 11, Quaternary Research, Science Council of Japan*, 16–34.

Yoshikawa, T., Kaizuka, S. and Ota, Y. (1981) 'Types and regional characteristics of Japanese plains'. In T. Yoshikawa, S. Kaizuka and Y. Ota *The Landforms of Japan*, University of Tokyo Press, Tokyo.

Young, M. and Bradley, R.S. 1984 'Insolation gradients and the palaeoclimatic record'. In A.L. Berger, J. Imbrie, J. Hays, G. Kukla and B. Saltzman (eds) *Milankovitch and Climate*, Part 2, 707–713, D. Reidel, Dordrecht.

INDEX

Northern Europe: glacial history 49–55
Northern hemisphere: cooling and dust
   veils in 181; ice age paleoclimates in
   31–8
Norway 222, 227, 250; deglaciation
   76–8; glacial history 51, 53, 55
Norwegian Sea: oxygen isotope
   stratigraphy from ocean sediments
   13–16

Ob River 48, 74, 157
ocean bottom water: role of in oxygen
   isotope stratigraphy 11–12
ocean sediments: and ice cores 6–23;
   Late Quaternary oxygen isotope
   stratigraphy 13–16; limitations of
   analysis of 12–13; oxygen isotopes in
   6–16
oceans: aeolian activity, evidence from
   176–8; circulation in Antarctic 39;
   general circulation models of 24;
   global thermohaline circulation cell
   34; ice age paleoclimates in Northern
   hemisphere 32–6; North Atlantic
   deep water production (NADW) 34
Odderdade interstadial 50, 244
Oerel interstadial 51
Ojibway, Lake 98–9, 100–1; and ice
   unloading and shoreline deformation
   208
Old Crow tephra 191
Older Dryas Stadial 77, 82
Omoto, K. 66
Opdyke, N.D. 7, 10, 12, 33
orbital processes (Milankovitch)
   236–41; combined effects 239;
   eccentricity, changes in 237–8;
   equinoxes, precession of 237–9;
   inclination, changes in 237–8
Orombelli, G. 67
Orr, P.C. 135
Ota, Y. 213, 223, 225
Owens Lake 136
oxygen isotope stratigraphy 4; and
   aeolian activity 177; and
   chronostratigraphy of Late
   Quaternary 45; curves, interpretation
   of 9–11; and glaciation of North
   America 58; and global cooling 248,
   255; and global environmental
   change 22–3; ice age variations 8–9;
   of ice cores 16–21; and Lake Agassiz

93; and Late Quaternary climatic
   change 13–16, 243–5; limitations of
   analysis of ocean sediments 6–16;
   modern variations 6–8; ocean bottom
   water, role of 11–12; and pollen
   records 130; and sea level changes
   223–6; and superfloods 154; and
   tephra deposits 187–9, 198;
   uncertainty in 10–11, 19; variations,
   effects of ice sheets on 9

Pacific Ocean 223
Pacific plate 212
Pakistan: fossil dune fields 174
Palacios, O. 105
paleoclimates: and computer simulations
   24–41; inferences in periglacial
   environments 123–4
paleosols: and aeolian activity 162, 170;
   and Late Quaternary climatic change
   243–4
Palivere moraine 74
Pampas: fossil dune fields 175
Panamint Lake 136
Papua-New Guinea 213, 244
Parkin, D.W. 176
Paskoff, R.P. 67
Patagonia 67, 227
patterned ground: in periglacial
   environments 112
Patton, P.C. 145
Pee Dee Formation (South Carolina) 7
Peltier, W.R. 199, 200, 204, 214
Penck, A. 63, 143
Pennington, W. 82
periglacial environments 109–25;
   paleoclimatic inferences 123–4; see
   also permafrost
permafrost: climates in Late Valdai
   glaciation 117; in Europe 122–3;
   evolution 112–14; in North America
   118–22; reconstruction of, problems
   109–12; in Soviet Union 114–17; in
   Valdai/Sartan glaciation 112–13
Perrott, R.A. 133–4, 142
Persson, C. 183
Petit, R.J. 177
Pewe, T.L. 61–2, 109, 112, 119–21
pingos: in periglacial environments 110,
   120–1, 122
Piper, D.J.W. 61, 231, 245, 251
Pirazzoli, P. 213, 223